BURLEIGH DODDS SCIENCE: INSTANT INSIGHTS

NUMBER 62

T0200038

Managing bacterial diseases of poultry

burleigh dodds
SCIENCE PUBLISHING

Published by Burleigh Dodds Science Publishing Limited
82 High Street, Sawston, Cambridge CB22 3HJ, UK
www.bdspublishing.com

Burleigh Dodds Science Publishing, 1518 Walnut Street, Suite 900, Philadelphia, PA 19102-3406, USA

First published 2022 by Burleigh Dodds Science Publishing Limited
© Burleigh Dodds Science Publishing, 2023. All rights reserved.

British Library Cataloguing in Publication Data
A catalogue record for this book is available from the British Library

ISBN 978-1-80146-420-8 (Print)
ISBN 978-1-80146-421-5 (ePub)

DOI 10.19103/9781801464215

Typeset by Deanta Global Publishing Services, Dublin, Ireland

Contents

Series list

Chapter 1

Zoonoses affecting poultry: the case of *Campylobacter*

Tom J. Humphrey and Lisa K. Williams, Swansea University, UK

1 Introduction

There are several bacterial zoonoses that affect, and/or are associated with, poultry worldwide, including those caused by *Campylobacter, Salmonella, Escherichia coli* and *Listeria* spp. In the United Kingdom, during the 1980s, an outbreak of *Salmonella* linked to eggs made headlines; this led to most laying hens in the United Kingdom, reared under the Lion Code, to be vaccinated against *Salmonella*. This control measure led to a reduction in the number of cases in the human population seen in the United Kingdom. *Campylobacter* is now the leading cause of bacterial foodborne illness, with the majority of cases being attributed to the consumption of poultry and poultry products (Wilson et al. 2010; Hermans et al. 2011; Levin 2007; Williams 2014), although there have been cases associated with pigs (Denis et al. 2008; Horrocks et al. 2009; Zweifel et al. 2008; Little et al. 2008; Oporto et al. 2007), cows (Gilpin et al. 2008; Kwan et al. 2008; Ingilis et al. 2004; Busato et al. 1999), sheep (Sahin et al. 2008; Oporto et al. 2007; Zweifel et al. 2004; Açik and Cetinkaya 2006), companion animals (Kemp et al. 2005; Brown et al. 2004; Andrzejewska et al. 2013) and wild birds (Horrocks et al. 2009; Kwan et al. 2008; Meerburg et al. 2006).

Despite stringent biosecurity on farm and control measures during processing, the number of cases of *Campylobacter* continues to rise each year. In the United Kingdom, it is thought that 82% of hospital admissions with a diagnosis of food poisoning can be attributed to a *Campylobacter* infection (Adak et al. 2002). In humans, the infectious dose is believed to be low, at around 500 cells in an adult male (Robinson 1981) and an incubation

http://dx.doi.org/10.19103/AS.2016.0010.01

period of up to 10 days, with most people exhibiting symptoms by day four. Symptoms include diarrhoea, which may be bloody, particularly in children; acute abdominal pain; and fever. Most infections are self-limiting and those who suffer an infection usually recover quickly after resting and maintenance of fluid levels; a fatal outcome is rare and would usually occur in the elderly or those already suffering from another serious illness (Skirrow and Blaser 2000). Despite treatment with antibiotics being rare, resistance to clinically important antimicrobials including macrolides and fluoroquinolones has been reported (Humphrey et al. 2007). Approximately 1% of cases develop long-term sequelae, such as inflammatory bowel syndrome or reactive arthritis.

2 *Campylobacter* in poultry

To date there are 26 species, 2 provisional species and 9 subspecies (Kaakoush et al. 2015) in the *Campylobacter* family. Of these species, *C. jejuni* and *C. coli* are most often found in the human cases, and are also commonly found in poultry meat (Humphrey et al. 2007).

Campylobacter has a long-standing association with poultry; it is well suited to 42°C, the body temperature of chickens, and is commonly found in the gastrointestinal tract of these birds (Hermans et al. 2011; Williams et al. 2014). *Campylobacter* was commonly referred to as a commensal of poultry (Dhillion et al. 2006; Macpherson and Uhr 2004) despite studies examining the impact of infection on the health and welfare of broiler chickens. In the field, symptoms we now know can be caused by *Campylobacter*, such as diarrhoea, were believed to be the action of other poultry pathogens. There have been studies based on laboratory and field data, going back to 1981, which have demonstrated that *Campylobacter* can and does have an effect on the health, welfare and performance of broiler chickens; this will be discussed in detail later.

There are two main routes of transmission from poultry to humans: first, via cross-contamination during processing, and secondly, via the spread of the bacterium from the intestines to other organs. Internationally, undercooked chicken muscle and liver are important vehicles for human infection. The chicken gut can be colonised with high numbers of *Campylobacter* spp. During processing, which is highly automated, these bacteria cross-contaminate the external surface of the carcass. *Campylobacter* can also be aerosolised, contaminating the carcass that way (Shane 1992; Gruntar et al. 2015; Elvers et al. 2011).

More recently, it has been shown that *Campylobacter* has the ability to leave the gut and infect other organs, the liver being the predominant one. Extra-intestinal spread is a major public health issue: several outbreaks of human campylobacteriosis have been linked to the consumption of chicken livers (Fernandez and Pison 1996; Mederiros et al. 2008; Cox et al. 2006; Whyte et al. 2006; O'Leary et al. 2009; Simaluiza et al. 2015), indicating that perhaps extra-intestinal spread is more common than previously thought. The ability of *Campylobacter* to leave the gut is poorly understood, requiring further investigation. *Campylobacter* spp. are isolated throughout poultry production, including rearing and at slaughter (Nebola et al. 2006), and their occurrence is well documented (Wagenaar et al. 2006; Humphrey 2006; Denis et al. 2001; Stern et al. 2001). However, little is known about the epidemiology of *Campylobacter* in poultry flocks (Bull et al. 2006), making control measures too difficult to implement and monitor.

3 Control measures in poultry

There is no doubt that controlling *Campylobacter* in poultry flocks would have an impact on the number of cases seen in humans; a vaccination strategy in the United Kingdom in the 1980s of laying flocks led to a reduction in the number of cases of *Salmonella* observed in the human population. Unfortunately, there is no effective vaccine for *Campylobacter* spp. for either the human or poultry populations, and this is mainly due to the complicated antigenic diversity *C. jejuni* has, a lack of small animal models and a poor understanding of *Campylobacter* pathogenesis (Riddle and Guerry 2016). In the absence of an effective vaccine, alternative control measures have been used during rearing on the farm and during processing with limited success.

During rearing, at the farm interventions including biosecurity, altering the diet, use of pre- and probiotics and additives (Meunier et al. 2015) have not been effective either individually or in combination. Other inventions during processing such as rapid chilling and surface washes have not been effective in reducing *Campylobacter* spp. to a level sufficient to lower the infection risk. *Campylobacter* has the ability to survive the harsh environments that it is exposed to during rearing and processing, and as a consequence of this, 71% of chicken carcasses on retail sale in the United Kingdom (Food Standards Agency 2015) contain *Campylobacter*. This is a problem not just in the United Kingdom; across Europe approximately 75% of carcasses on retail sale contain *Campylobacter* spp. Despite the interventions during rearing, processing and media campaigns aimed at raising awareness to consumers, the number of cases in the human population in the United Kingdom increases each year. There is no 'silver bullet' that will eradicate *Campylobacter* from poultry flocks, and individually, no single method appears to have a long-term effect on the bacteria. Combinations of control measures have been tried, including the use of antimicrobial feed additives containing fatty acids; pro- and prebiotics have been used together and combined with improved biosecurity (Meunier et al. 2015). However, in some cases these treatments predisposed birds to disease and increased the number of *Campylobacter*. Control measures that have proven to be effective in one country do not necessarily work in another. A good example of this is the use of fly screen on poultry houses in Denmark that led to a reduction in the level of *Campylobacter* in the flock (Hald et al. 2007, 2008), but when fly screens were used on poultry houses elsewhere, little effect was observed.

Other studies (Janez and Loc Carrillo 2013; Loc Carrillo et al. 2005; Atterbury et al. 2005) have examined the use of bacteriophage against *Campylobacter*. Although this has been shown in some experiments to be effective in controlling *Campylobacter*, its use is limited as consumers need convincing that these are 'good viruses' (Janez and Loc Carrillo 2013).

There are other control measures that have undergone testing, including rapid chilling, irradiation and use of elevated chlorine levels with carcasses. These have been tested with limited success (Berrang et al. 2011; James et al. 2007; Collins et al. 1996). Rapid chilling and irradiation have one major drawback: they affect carcass appearance and quality. Interventions are only useful if they are effective in controlling *Campylobacter* and are reported to have no impact on carcass appearance or meat quality. There are also some concerns that consumers do not find irradiation acceptable. There are strict controls in the United Kingdom regulating the amount of chlorine that is permitted to be used in potable water; and so, although chlorine has been shown to be effective (Berrang et al. 2011) in reducing *Campylobacter* levels during processing, the level used cannot be greater than is permitted by regulations.

4 *Campylobacter* as a pathogen rather than a commensal of poultry

Whether *Campylobacter* is a pathogen or a commensal of poultry is a topic of debate in the wider *Campylobacter* research community. Historically, *Campylobacter* was referred to as a commensal of poultry, as it was believed to cause no obvious signs of disease or have an effect on the health and welfare of broiler chickens. *Campylobacter* is considered part of the caecal microbiome of broiler chickens, because it is found in nearly every flock reared in the United Kingdom. It is generally accepted that every broiler chicken will be colonised with or exposed to *Campylobacter* during their lifetime irrespective of breed and rearing system used. Birds usually become colonised by the time they are three weeks old, and due to the rapid growth and short lifespan of the birds, the immune system is underdeveloped, which may make them more vulnerable to infection. However, chickens do mount innate and adaptive immune responses to infection with *C. jejuni* (Humphrey et al. 2014), but this does not result in clearance and may have the objective of confining the bacteria to the gut.

In many ways *Campylobacter* in broiler chickens can meet the criteria to be considered a commensal, although this can depend on the strain of the bacterium and broiler type (Humphrey et al. 2014). However, the reason for this is that, historically, signs of disease, which we now know can be associated with *Campylobacter*, were thought to be due to other bacteria or diseases. There have been changes in the performance and welfare of broiler chickens that are associated with other conditions and with carcass rejections at slaughter, which are examined thoroughly. For more than 30 years *Campylobacter* had been considered part of the normal gut microbiota of a chicken, despite early evidence indicating otherwise (Neil et al. 1984; Ruiz-Palacious et al. 1981; Sanyal et al. 1984). Certainly within some bird types and with certain strains of *Campylobacter* the behaviour of the bacterium is more like a commensal, but this is not universal.

It was only recently accepted that chickens mount an immune response to *Campylobacter*. Prior to this, chickens were thought not to do so in the same way as they would not for other commensal pathogens such as *Lactobacillus*. As we develop a deeper understanding of the chicken immune system, we begin to realise that it is more complex than previously thought and that *Campylobacter* is dealt with by the chicken immune system in a similar way to other pathogens. Many still consider *Campylobacter* to be a commensal of chickens, but there is an increasing body of evidence from several research groups (Awad et al. 2014; Williams et al. 2013, 2014; Wigley and Humphrey 2014; Humphrey et al. 2014), suggesting that it can no longer be considered in this way.

Several studies have reported the ability of *Campylobacter* to leave the chicken gut and infect other internal organs, including the liver (Knudsen et al. 2006; Van Deun et al. 2007; Sanyal et al. 1984; Jennings et al. 2011; O'Leary et al. 2009). Muscle contamination has also been reported in some studies (Berntson et al. 1992; Scherer et al. 2006; Luber and Bartlett 2007). The source of this contamination has previously been debated, and cross-contamination from the intestine during processing is not the only contributing factor to infected meat (Williams et al. 2014). Studies dating back to 1984 (Sanyal et al.1984; Knudsen et al. 2006; O'Leary et al. 2009; Jennings et al. 2011) have isolated *Campylobacter* from organs other than the gut; if *Campylobacter* was a true gut commensal of chickens, it should only be found there.

It is well understood that, within the poultry gut, *Campylobacter* adheres to epithelial cells and that this adherence is essential for colonisation (Hermans et al. 2011). Extra-intestinal spread does not always indicate disease or that an infectious process is occurring (Williams et al. 2014), but the invasive behaviour of *Campylobacter* in chickens suggests that it should be regarded as either a true or an opportunistic pathogen. It is increasingly being recognised that *Campylobacter* does alter the state of the host (Williams et al. 2014; Humphrey et al. 2014).

Campylobacter can also have an effect on the health, welfare and performance of broiler chickens, and there has been an association between certain welfare-associated conditions such as hock marks and pododermatitis. These foot and leg lesions are associated with diarrhoea in the birds and are considered risk factors for *Campylobacter* infection in commercial flocks (Neil et al. 1984; Bull et al. 2008; Rushton et al. 2009). Williams et al. (2013) demonstrated that the incidence and severity of hock marks and pododermatitis increased when *Campylobacter* is present in artificially infected birds. These leg and foot conditions are caused by prolonged contact with wet litter, and this study strongly suggests that *Campylobacter* is causing these leg lesions indirectly by increasing the looseness of the faeces, which increases the wetness of the litter (Williams et al. 2013, 2014). Similar results were obtained by Humphrey et al. (2014), who showed that diarrhoea in certain broiler breeds was associated with high levels of inflammation and much damage to gut mucosa (Humphrey et al. 2014; Fig. 1). Figure 1 is taken from Humphrey et al. (2014) and shows the breed with high levels of diarrhoea and raised levels of pododermatitis, which exhibited profound damage to the mucosa of the ileum, whereas in the breed with essentially normal faeces and no pododermatitis, the damage was much less (Fig. 1).

Damage to gut mucosa has been seen in other studies using artificial infection of the rapid growing breeds currently used commercially (Awad et al. 2014 and 2015). These studies provide evidence to suggest that *Campylobacter* is having a direct impact on the health and welfare of broiler chickens. Awad et al. (2014, 2015) showed that birds infected with *Campylobacter* performed significantly less well in the laboratory. Ruiz-Palacios

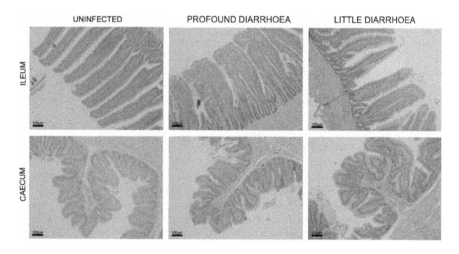

Figure 1 In some broilers *C.jejuni* damages the gut epithelia.

et al. (1981) showed that, in chickens infected with *C. jejuni* isolates from human cases of diarrhoea, over 80% of the birds got diarrhoea and around 40% died; and there was around a 40% drop in growth rate in the survivors, making this the first documented evidence suggesting that *Campylobacter* has an effect on the health and welfare of broiler chickens. This has previously been overlooked, allowing *Campylobacter* to be referred to as a commensal when the evidence indicates otherwise. A recent paper from India (Singh et al. 2012) found that the presence of *Campylobacter* in the gut microbiota of chickens was significantly ($p < 0.0004$) associated with poorer feed conversion ratios (FCRs).

Recent fieldwork in the United Kingdom and Ireland (Sparks and Whyte pers. comm.) has found that *Campylobacter*-positive commercial broiler flocks had significantly reduced performance than negative ones, although there were bacterial strain-to-strain differences in the impact of *Campylobacter*. These studies suggest that there are both economic and welfare reasons for better *Campylobacter* control on farm; these bacteria threaten the economic sustainability of chicken production, and the infection of chickens is likely to cost the international poultry industry considerable a lot each year.

Campylobacter being linked to disease in the chicken is not new; in 1954, a link was made between vibrio-like organisms, which were later identified as *C. jejuni*, and vibrionic hepatitis (Tudor 1954). Other authors have found an association between avian vibrio hepatitis (AVH) and the presence of *Campylobacter* (Lukas 1955; Hofstad et al. 1958; Moore 1958; Sevolan et al. 1958; Whenham et al. 1961). This disease, which causes focal lesions that are greyish-white in colour and 1–2 mm in size (Burch 2005; Crawshaw and Young 2003), persisted until 1965 and re-emerged in the 2000s (Crawshaw and Young 2003; VLA surveillance report). *Campylobacter* has not been conclusively linked with the focal lesions (Jennings et al. 2011) despite being initially linked with the disease (Peckham 1958). Jennings et al. (2011) found that significantly fewer healthy livers contained *Campylobacter* than those with greyish white spots, and using FISH they demonstrated that *Campylobacter* was present in higher numbers in spotty livers than in livers without spots, which were deemed healthy. *Campylobacter* has been found previously in the livers of apparently healthy birds (Cox et al. 2006, 2007, 2009). A further study (Richardson et al. 2011) identified *Campylobacter* in the bloodstream of broiler chickens with blood taken from naturally infected birds at slaughter, as did Sanyal et al. (1984), using artificially infected birds.

As the bacteria may be present in the blood of birds without apparently causing harm to these birds, and thus no rejection at slaughter, they could lodge in the small blood vessels in chicken muscle when the birds are bled. Undercooked chicken meat is an internationally important vehicle for human infection. In their experimental studies, Ruiz-Palacios et al. (1981) found that around 15% of birds had *Campylobacter* in circulating blood, and Richardson et al. (2011) found approximately 12%.

Damage to gut mucosa has been reported in some studies (Gharib et al. 2012; Awad et al. 2014; Ruiz-Palacios et al. 1981; Sanyal et al. 1984; Humphrey et al. 2014), but not all (Dhillion et al. 2006; Larson et al. 2008). There were also differences in whether birds suffered diarrhoea. Some authors reported this (Humphrey et al. 2014; Sanyal 1984, Ruiz-Palacios et al.1981; Sang 1989), particularly in young chickens. The bird type could have an impact on the effects of *Campylobacter*, but bacterial strain is also important. The same type of bird has been used in several studies, but has been colonised with different *C. jejuni* strains (Williams et al. 2013; Awad et al. 2014; Humphrey et al. 2014), and diarrhoea was not reported in all. Differences between strains in their ability to cause diarrhoea in chickens have been reported by others (Humphrey et al. 2014). It is clear that certain *Campylobacter* strains cause disease in different chicken types, although few papers have looked beyond caecal colonisation.

The chicken immune system is highly developed, but in comparison with mammals, it is considered smaller and simpler, although the functions are the same. The immune response to *Campylobacter* colonisation is complex, and little is known about the interaction between *Campylobacter* and the chicken immune system (Wigley and Humphrey 2014). *Campylobacter* is recognised by pattern-recognition receptors (PRRs), which lead to the activation of an immune response. Toll-like receptors (TLRs), in particular, in terms of *Campylobacter* colonisation, TLR 4 and TLR 21, are activated in the gut, although it has been reported that the increase in the expression of these TLRs in response to *Campylobacter* colonisation is short-lived (Shaughnessy et al. 2009). Following PRR activation, there is an increase in the pro-inflammatory cytokines, IL-6 and IL-1β, along with a pro-inflammatory chicken chemokine IL-8 (chIL-8) (Larson 2008; Shaughnessy 2000). After this initial pro-inflammatory response, cytokine levels decrease, suggesting that, although initial colonisation is dealt with by the host immune system as an attack, it reaches a level of tolerance (Williams et al. 2014; Smith et al. 2008; Hermans et al. 2012). Previous studies (Poh 2008) have shown that both hetrophils and monocytes are attracted by IL-8 to areas of inflammation. However, there is conflicting evidence whether this happens in IL-8 produced with *Campylobacter* colonisation (Williams et al. 2014). Several studies have shown no heterophilia (Meade 2009; Van Deun 2008), but Smith et al. (2008) observed an increase in hetrophils in the caeca and ileum of experimentally infected birds at two weeks of age. TLR5 recognises flagellin and there is a strong inflammatory response to invasion, mainly mediated by the recognition of flagellin through TLR5, leading to the production of CXCL chemokines and pro-inflammatory cytokines including IL-6 and IL-1β (Iqbal et al. 2005; Wigley 2013). This leads to inflammatory damage, which is considered to be mild in some chicken types (Humphrey et al. 2014), and can also lead to immune activation and largely restricts infection to the gut (Henderson et al. 1999; Withanage et al. 2004; Wigley 2013). Certain strains of *Salmonella* have adapted to cause systemic infection in the chicken, namely the serovars Gallinarum and Pullorum lack flagella and so evade recognition by the host immune system, allowing systemic infection to occur (Chappell et al. 2009). In a similar way to *Salmonella* Gallinarum and Pullorum, *Campylobacter* can also evade TLR-5 recognition by glycosylation of its flagella (Howard et al. 2009), but it is recognised by other receptors including TLR-21 (Meade et al. 2009). Several studies (Humphrey et al. 2014; Meade et al. 2009; Shaughnesaay et al. 2009; Smith et al. 2008) have shown that there is an inflammatory response to *Campylobacter* in the intestine. Although the bacterium was believed to be poorly invasive, it has been found in other organs, including the liver (Knudsen et al. 2006; Van Deun et al. 2007; Sanyal et al. 1984; Jennings et al. 2011; O'Leary et al. 2009). Infection of the liver can also be influenced by the *Campylobacter* strain (Medeiros et al. 2008; Cox et al. 2006; Whyte et al. 2006; O'Leary et al. 2009).

As well as mounting an innate immune response to *Campylobacter*, chickens also mount an adaptive one, producing antibodies against a number of *Campylobacter* proteins, including flagellin (Cawthraw et al. 1994; Sahin et al. 2001, 2002; Smith et al. 2008). Our understanding of the adaptive response of chicken to *Campylobacter* is rudimentary (Wigley and Humphrey 2014). Chickens have been shown to produce a humoral response against T-cell-dependent and T-cell-independent type 1 antigens; they produce a low response against T-cell-independent type 2 antigens (Jeurissen et al. 1998). The polysaccharide component of the *Campylobacter* capsule is a type 2 antigen and this has been suggested to contribute to the lack of clearance of *Campylobacter* from the chicken gut (Williams et al. 2014).

There is no doubt that chickens mount a response to infection; although an initial one is to be expected to any commensal organism, reports of pathology including heterophilia indicate that *Campylobacter* may be causing a diseased-like state, and is behaving more like a pathogen than a commensal (Williams et al. 2014; Humphrey et al. 2015).

5 Conclusions

There is strong evidence that indicates that *Campylobacter* from broiler chickens causes the majority of cases of human bacterial foodborne illness. Despite interventions on the farm during rearing and during processing the incidence of *Campylobacter* does not seem to be decreasing, indicating a need for an alternative approach. There is a need to understand the biology of *Campylobacter* in the chicken to determine the effect it is having on the health, welfare and performance of broiler chickens. Historically, *Campylobacter* has been referred to as a commensal of broiler chickens, but there is increasing evidence to challenge this viewpoint. Several studies, from different research groups, have shown that *Campylobacter* is capable of invading the gut cells, causing extra-intestinal infection of other internal organs and muscle. In addition, it has been shown that chickens mount innate and adaptive immune responses to *Campylobacter* higher than they do to commensal bacteria. The evidence suggests that, under the right conditions, *Campylobacter* behaves more like a pathogen, which can have a direct effect on the health and welfare of chickens; therefore, it should not always be considered a commensal of chickens.

6 Future trends

Campylobacter is an ongoing problem to the food industry. Chicken is the main reservoir for human infections, and the number of cases in the human population has largely increased despite interventions on the farm and during processing. There is no 'silver bullet' to tackle *Campylobacter*; it is likely to be a combination of factors that will lead to significant reductions in the number of cases observed in the human population. Ongoing work is looking at the viability of a vaccine for *Campylobacter* and work in this area will continue. There is a need to understand the effect *Campylobacter* has on a broiler chicken; for so long it has been considered a harmless commensal despite evidence to challenge this view. It is likely that the high incidence of human infection by *Campylobacter* will continue in most countries worldwide until we fully understand the infection biology of different strains of these bacteria in various broiler chicken breeds.

7 Where to look for further information

A good review on *Campylobacter* being more than a commensal in chickens is Williams et al. (2014) Conditional Commensalism of *Campylobacter* in chickens, which is in the book *Campylobacter* Ecology and Evolution Edited by S. K. Sheppard. There are also chapters in this book on the chicken immune system and responses to *Campylobacter*.

The World Health Organization is a useful resource for detailed information on *Campylobacter*, http://www.who.int/topics/campylobacter/en/

The European Food Standards Agency regularly publishes advice, trends and information about *Campylobacter* and poultry, https://www.efsa.europa.eu/en/topics/topic/campylobacter

Individual countries also publish their own guidance, trends and information. In the UK the Food Standards Agency, https://www.food.gov.uk/science/microbiology/campylobacterevidenceprogramme and Public Health England, https://www.gov.uk/government/collections/campylobacter-guidance-data-and-analysis are good sources of information, particularly about human campylobacterosis.

In the US the Centers for Disease Control and Prevention (CDC), http://www.cdc.gov/foodsafety/diseases/campylobacter/index.html and the US Food and Drug Administration http://www.fda.gov/default.htm contain valuable information.

Details on how UK funding bodies have joined together to tackle *Campylobacter* in the food chain can be found here, http://www.bbsrc.ac.uk/innovation/collaboration/collaborative-programmes/tackling-campylobacter/

Campylobacter in poultry information can be found here http://www.thepoultrysite.com/diseaseinfo/22/campylobacter-infection/

8 References

Açik, M. N. and Cetinkaya, B. (2006). Random amplified polymorphic DNA analysis of *Campylobacter jejuni* and *Campylobacter coli* isolated from healthy cattle and sheep. *J. Med. Microbiol.* 55(Pt 3): 331–4.

Adak, G. K., Long, S. M. and O'Brien, S. J. (2002). Trends in indigenous foodborne disease and deaths, England and Wales: 1992–2002. *Gut* 51: 832–41.

Andrzejewska, M., Szczepanska, B., Klawe, J. J., Spica, D. and Chudzinska, M. (2013). Prevalence of *Campylobacter jejuni* and *Campylobacter coli* species in cats and dogs from Bydgoszcz (Poland) region. *Polish J. Vet. Sci.* 1: 115–20.

Atterbury, R. J., Dillon, E., Swift, C., Connerton, P. L., Frost, J. A., Dodd, C. E. R., Rees, C. E. D. and Connerton, I. F. (2005). Correlation of *Campylobacter* bacteriophage with reduced presence of hosts in broiler chicken ceca. *Appl. Environ. Microbiol.* 71: 4885–7.

Awad, W. A., Aschenbach, J. R., Chareeb, K., Khayal, B., Hess, C. and Hess, M. (2014). *Campylobacter jejuni* influences the expression of nutrient transporter genes in the intestines of chickens. *Vet. Microbiol.* 172: 195–201.

Awad, W. A., Molnár, A., Aschenbach, J. R., Chareeb, K., Khayal, B., Hess, C., Liebhart, D., Dublecz, K. and Hess, M. (2015). *Campylobacter infection* in chickens modulates the intestinal epithelial barrier function. *Innate Immun.* 19: 1–10.

Berndtson, E., Tiverno, M. and Engvall, A. (1992). Distribution and numbers of *Campylobacter* in newly slaughtered broiler chickens and hens. *Int. J. Food Microbiol.* 15: 45–50.

Berrang, M. E., Windham, W. R. and Meinersmann, R. J. (2011). *Campylobacter, Salmonella,* and *Escherichia coli* on broiler carcasses subjected to a high pH scald and low pH postpick chlorine dip. *Poult. Sci.* 90: 896–900.

Brown, P. E., Christensen, O. F., Clough, H. E., Diggle, P. J., Hart, C. A., Hazel, S., Kemp, R., Leatherbarrow, A. J. H., Moore, A., Sutherst, J., Turner, J., Williams, N. J., Wright, E. J. and

French, N. P. (2004). Frequency and spatial distribution of environmental *Campylobacter* spp. *Appl. Environ. Microbiol.* 70: 6501–11.

Bull, S. A., Thomas, A., Humphrey, T., Ellis-Iversen, J., Cook, A. J., Lovell, R. and Jorgensen, F. (2008). Flock health indicators and *Campylobacter* spp. in commercial housed broilers reared in Great Britain. *Appl. Environ. Microbiol.* 74: 5408–13.

Burch, D. (2005). Avian vibrionic hepatitis in laying hens. *Vet. Rec.* 157: 528.

Busato, A., Hofer, D., Lentze, T., Gaillard, C. and Burnens, A. (1999). Prevalence and infection risks of zoonotic enteropathogenic bacteria in Swiss cow-calf farms. *Vet. Microbiol.* 69: 251–63.

Cawthraw, S., Ayling, R., Nuijten, P., Wassenaar, T. and Newell, D. G. (1994). Isotype, specificity, and kinetics of systemic and mucosal antibodies to *Campylobacter jejuni* antigens, including flagellin, during experimental oral infections of chickens. *Avian. Dis.* 38: 341–9.

Chappell, L., Kaiser, P., Barrow, P., Jones, M. A., Johnston, C. and Wigley, P. (2009). The immunobiology of avian systemic salmonellosis. *Vet. Immunol. Immunopathol.* 128: 53–9.

Collins, C. L., Murano, E. A. and Wesley, I. V. (1996). Survival of *Arcobacter butzleri* and *Campylobacter jejuni* after irradiation treatment in vacuum-packaged ground pork. *J. Food Protect.* 11: 1153–247.

Cox, N. A., Richardson, L. J., Buhr, J. R., Fedorka-Cray, J. P., Bailey, J. S., Wilson, J. L. and Hiett, K. L. (2006). Natural presence of *Campylobacter* spp. in various internal organs of commercial broiler breeder hens. *Avian. Dis.* 50: 450–3.

Cox, N. A., Richardson, L. J., Buhr, R. J. and Fedorka-Cray, P. J. (2009). *Campylobacter* species occurrence within internal organs and tissues of commercial caged Leghorn laying hens. *Poult. Sci.* 88: 2449–56.

Cox, N. A., Richardson, L. J., Buhr, R. J., Northcutt, J. K., Bailey, J. S. and Fedorka-Cray, P. J. (2007). Recovery of *Campylobacter* and *Salmonella* serovars from the spleen, liver/gallbladder, and ceca of six and eight week old commercial broilers. *J. Appl. Poult. Res.* 16: 477–80.

Crawshaw, T. and Young, S. (2003). Increased mortality on a free-range layer site. *Vet. Rec.* 153: 664.

Denis, M., Refregier-Petton, J., Laisney, M. J., Ermel, G. and Salvat, G. (2001). *Campylobacter* contamination in French chicken production from farm to consumers. Use of a PCR assay for detection and identification of *C. jejuni* and *C. coli. J. App. Microbio.* 91: 255–67.

Denis, M., Rose, V., Huneau-Salaün, A., Balaine, L. and Salvat, G. (2008). Diversity of pulsed-field gel electrophoresis profiles of *Campylobacter jejuni* and *Campylobacter coli* from broiler chickens in France. *Poult. Sci.* 87: 1662–71.

Dhillon, A. S., Shivaprasad, H. L., Schaberg, D., Wier, F., Weber, S. and Bandli, D. (2006). *Campylobacter jejuni* infection in broiler chickens. *Avian Dis.* 50: 55–8.

Elvers, K. T., Morris, V. K., Newell, D. G. and Allen, V. M. (2011). Molecular tracking, through processing, of *Campylobacter* strains colonizing broiler flocks. *Appl. Envion. Microbiol.* 77: 5722–9.

Fernández, H. and Pisón, V. (1996). Isolation of enteropathogenic thermotolerant species of *Campylobacter* from commercial chicken livers. *Int. J. Food Microbiol.* 29: 75–80.

Food Standards Agency (2015). *Campylobacter* retail survey. Available from, http://www.food.gov.uk/news-updates/campaigns/campylobacter/actnow/act-e-newsletter/campylobacter-retail-survey last (accessed 04 April 2016).

Gharib Naseri, K., Rahimi, S. and Khaki, P. (2012). Comparison of the effects of probiotic, organic acid and medicinal plant on *Campylobacter jejuni* challenged broiler chickens. *J. Agr. Sci. Tech.* 14: 1485–96.

Gilpin, B. J., Thorrold, B., Scholes, P., Longhurst, R. D., Devane, M., Nicol, C., Walker, S., Robson, B. and Savill, M. (2008). Comparison of *Campylobacter jejuni* genotypes from dairy cattle and human sources from the Matamata-Piako District of New Zealand. *J. Appl. Microbiol.* 105: 1354–60.

Gruntar, I., Biasizzo, M., Kusar, D., Pate, M. and Ocepek, M. (2015). *Campylobacter jejuni* contamination of broiler carcasses: population dynamics and genetic profiles at slaughterhouse level. *Food Microbiol.* 50: 97–101.

Hald, B., Skovgård, H., Pedersen K. and Bunkenborg, H. (2008). Influxed insects as vectors for *Campylobacter jejuni* and *Campylobacter coli* in Danish broiler houses. *Poult. Sci.* 87: 1428–34.

Hald, B., Sommer, H. M. and Skovgård, H. (2007). Use of fly screens to reduce *Campylobacter* spp. introduction in broiler houses. *Emer. Infect. Dis.* 13: 1951–3.

Hermans, D., Pasmans, F., Heyndrickx, M., Van Immerseel, F., Martel, A., Van Deun, H. and Haesebrouck, F. (2012). A tolerogenic mucosal immune response leads to persistent *Campylobacter jejuni* colonization in the chicken gut. *Crit. Rev. Microbiol.* 38: 17–29.

Hermans, D., Van Deun, K., Martel, A., Van Immerseel, F., Messens, W., Heyndrickx, M., Haesebrouck, F. and Pasmans, F. (2011). Colonization factors of *Campylobacter jejuni* in the chicken gut. *Vet. Res.* 42: 82.

Hofstad, M. S., McGehee, E. H. and Bennett, P. C. (1958). Avian infectious hepatitis. *Avian Dis.* 2: 358.

Horrocks, S. M., Anderson, R. C., Nisbet, D. J. and Ricke, S. C. (2009). Incidence and ecology of *Campylobacter jejuni* and *coli* in animals. *Anaerobe* 15: 18–25.

Humphrey, S., Chaloner, G., Kemmett, K., Davidson, N., Williams, N., Kipar, A., Humphrey, T. and Wigley, P. (2014). *Campylobacter jejuni* is not merely a commensal in commercial broiler chickens and affects bird welfare. *MBio* 5: 01364.

Humphrey, S., Lacharme-Lora, L., Chaloner, G., Gibbs, K., Humphrey, T., Williams, N. and Wigley, P. (2015). Heterogeneity in the infection biology of *Campylobacter jejuni* isolates in three infection models reveals an invasive and virulent phenotype in a ST21 isolate from poultry. *PLoS ONE*. 2310: e0141182.

Humphrey, T. (2006). Are happy chickens safer chickens? Poultry welfare and disease susceptibility. *Br. Poult. Sci.* 47: 379–91.

Humphrey, T. J., O'Brien, S. and Madsen, M. (2007). Campylobacters as zoonotic pathogens: a food production perspective. *Int. J. Food Microbiol.* 117: 237–57.

Inglis, G. D., Kalischuk, L. D. and Busz, H. W. (2004). Chronic shedding of *Campylobacter* species in beef cattle. *J. Appl. Microbiol.* 97: 410–20.

Iqbal, M., Philbin, V. J., Withanage, G. S., Wigley, P., Beal, R. K., Goodchild, M. J., Barrow, P., McConnell, I., Maskell, D. J., Young, J., Bumstead, N., Boyd, Y. and Smith, A. L. (2005). Identification and functional characterization of chicken toll like receptor 5 reveals a fundamental role in the biology of infection with *Salmonella enterica* serovar *typhimurium*. *Infect. Immun.* 73: 2344–50.

James, C., James, S. J., Hannay, N., Purnell, G., Barbedo-Pinto, C., Yaman, H., Araujo, M., Luisa Gonzalex, M., Calvo, J., Howell, M. and Corry, J. E. L. (2007). Decontamination of poultry carcasses using steam or hot water in combination with rapid cooling, chilling or freezing of carcass surfaces. *Int. J. Food. Microbiol.* 114: 195–203.

Janez, N. and Loc-Carrillo, C. (2013). Use of phages to control *Campylobacter* spp. *J. Micro. Methods* 95: 68–75.

Jennings, J. L., Sait, L. C., Perrett, C. A., Foster, C., Williams, L. K., Humphrey, T. J. and Cogan, T. A. (2011). *Campylobacter jejuni* is associated with, but not sufficient to cause vibrionic hepatitis in chickens. *Vet. Microbiol.* 149: 193–9.

Jeurissen, S. H., Janse, E. M., van Rooijen, N. and Claassen, E. (1998). Inadequate anti-polysaccharide antibody responses in the chicken. *Immunobiology* 198: 385–95.

Kaakoush, N. O., Castano-Rodriguez, N., Mitchell, H. M. and Man, S. I. M. (2015). Global epidemiology of *Campylobacter* infection. *Clin. Microbiol. Rev.* 28: 687–720.

Kemp, R., Leatherbarrow, A. J. H., Williams, N. J., Hart, C. A., Clough, H. E., Turner, J., Wright, E. J. and French, N. P. (2005). Prevalence and genetic diversity of *Campylobacter* spp. in environmental water samples from a 100-square-kilometer predominantly dairy farming area. *Appl. Environ. Microbiol* 71: 1876–82.

Knudsen, K. N., Bang, D. D., Andresen, L. O. and Madsen, M. (2006). *Campylobacter jejuni* strains of human and chicken origin are invasive in chickens after oral challenge. *Avian Dis.* 50: 10–14.

Kwan, P. S., Barrigas, M., Bolton, F. J., French, N. P., Gowland, P., Kemp, R., Leatherbarrow, H., Upton, M. and Fox, A. J. (2008). Molecular epidemiology of *Campylobacter jejuni* populations in dairy cattle, wildlife, and the environment in a farmland area. *Appl. Environmen. Microbiol.* 74: 5130–8.

Larson, C. L., Shah, D. H., Dhillon, A. S., Call, D. R., Ahn, S., Haldorson, G. J., Davitt, C. and Konkel, M. E. (2008). *Campylobacter jejuni* invade chicken LMH cells inefficiently and stimulate differential expression of the chicken CXCLi1 and CXCLi2 cytokines. *Microbiology* 154: 3835–47.

Levin, R. E. (2007). *Campylobacter jejuni*: a review of its characteristics, pathogenicity, ecology, distribution, subspecies characterization and molecular methods of detection. *Food Biotechnol.* 21: 271–347.

Little, C. L., Richardson, J. F., Owen, R. J., de Pinna, E. and Threlfall, E. J. (2008). Prevalence, characterisation and antimicrobial resistance of *Campylobacter* and *Salmonella* in raw poultrymeat in the UK, 2003–5. *Int. J. Environ. Health Res.* 18: 403–14.

Loc Carillo, C., Atterbury, R. J., El-Shibiny, A., Connerton, P. L., Dillon, E., Scott, A. and Connerton, I. F. (2005). Bacteriophage therapy to reduce *Campylobacter jejuni* colonization of broiler chickens. *Appl. Environ. Microbiol.* 71: 6554–63.

Luber, P. and Bartelt, E. (2007). Enumeration of *Campylobacter* spp. on the surface and within chicken breast fillets. *J. Appl. Microbiol.* 102: 313–8.

Lukas, G. N. (1955). Avian infectious hepatitis-a preliminary report. *J. Am. Vet. Med. Assoc.* 126: 402.

Macpherson, A. J. and Uhr, T. (2004). Compartmentalization of the mucosal immune responses to commensal intestinal bacteria. *Ann. NY Acad. Sci.* 1029: 36–43.

Meade, K. G., Narciandi, F., Cahalane, S., Reiman, C., Allan, B. and O'Farrelly, C. (2009). Comparative in vivo infection models yield insights on early host immune response to *Campylobacter* in chickens. *Immunogenetics* 61: 101–10.

Medeiros, D. T., Sattar, S. A., Farber, J. M. and Carrillo, C. D. (2008) Occurrence of *Campylobacter* spp. in raw and ready-to-eat foods and in a Canadian food service operation. *J. Food Prot.* 71: 2087–93.

Meerburg, B. G., Jacobs-Reitsma, W. F, Wagenaar, J. A. and Kijlstra, A. (2006). Presence of *Salmonella* and *Campylobacter* spp. in wild small mammals on organic farms. *Appl. Environ. Microbiol.* 72: 960–2.

Meunier, M., Guyard-Nicodème, M., Dory, D. and Chemaly, M. (2015). Control strategies against *Campylobacter* at the poultry production level: biosecurity measures, feed additives and vaccination. *J. Appl. Microbiol.* (Epub ahead of print).

Moore, R. W. (1958). Studies of an agent causing hepatitis in chickens. *Avian Dis.* 2: 39.

Nebola, M. and Steinhauserova, I. (2006). PFGE and PCR/RFLP typing of *Campylobacter jejuni* strains from poultry. *Br. Poult. Sci.* 47: 456–61.

Neil, S. D., Campbell, J. N. and Greene, J. A. (1984). *Campylobacter* species in broiler chickens. *Avian Pathol.* 13: 777–85.

O'Leary, M. C., Harding, O., Fisher, L. and Cowden, J. (2009). A continuous common-source outbreak of campylobacteriosis associated with changes to the preparation of chicken liver pâté. *Epidemiol. Infect.* 137: 383–8.

Oporto, B., Esteban, J. I., Aduriz, G., Juste, R. A. and Hurtado, A. (2007). Prevalence and strain diversity of thermophilic *Campylobacters* in cattle, sheep and swine farms. *J. Appl. Microbiol.* 103: 977–84.

Peckham, M. C. (1958). Avian vibrionic hepatitis. *Avian Dis.* 2: 348.

Poh, T. Y., Pease, J., Young, J. R., Bumstead, N. and Kaiser, P. (2008). Re-evaluation of chicken CXCR1 determines the true gene structure: CXCLi1 (K60) and CXCLi2 (CAF/interleukin-8) are ligands for this receptor. *J. Biol. Chem.* 283: 16408–15.

Richardson, L. J., Cox, N. A., Buhr, R. J. and Harrison, M. A. (2011). Isolation of *Campylobacter* from circulating blood of commercial broilers. *Avian Dis.* 55(3): 375–8.

Riddle, M. S. and Guerry, P. (2016). Status of vaccine research and development for *Campylobacter jejuni*. *Vaccine* (in press).

Robinson, D. A. (1981). Infective dose of *Campylobacter jejuni* in milk. *Br. Med. J.* 282: 1584.

Ruiz-Palacios, G. M, Escamilla, E. and Torres, N. (1981). Experimental *Campylobacter* diarrhea in chickens. *Infect. Immun.* 34: 250–5.

Rushton, S. P., Humphrey, T. J., Shirley, M. B., Bull, S. and Jorgensen, F. (2009). *Campylobacter* in housed broiler chickens: a longitudinal study of risk factors. *Epidemiol. Infect.* 37: 1099–110.

Sahin, O., Zhang, Q., Meitzler, J. C., Harr, B. S., Morishita, T. Y. and Mohan, R. (2001). Prevalence, antigenic specificity, and bactericidal activity of poultry anti-*Campylobacter* maternal antibodies. *Appl. Environ. Microbiol.* 67: 3951–7.

Sahin, O., Morishita, T. Y. and Zhang, Q. (2002). *Campylobacter* colonization in poultry: sources of infection and modes of transmission. *Anim. Health Res. Rev.* 3: 95–105.

Sahin, O., Plummer, P. J., Jordan, D. M., Sulaj, K., Pereira, S., Robbe-Austerman, S., Wang, L., Yaeger, M. J., Hoffman, L. J. and Zhang, Q. (2008). Emergence of a tetracycline-resistant *Campylobacter jejuni* clone associated with outbreaks of ovine abortion in the United States. *J. Clin. Microbiol.* 46: 1663–71.

Sang, F. C., Shane, S. M., Yogasundram, K., Hagstad, H. V. and Kearney, M. T. (1989). Enhancement of *Campylobacter jejuni* virulence by serial passage in chicks. *Avian Dis.* 33: 425–30.

Sanyal, S. C., Islam, K. M., Neogy, P. K., Islam, M., Speelman, P. and Huq, M. I. (1984). *Campylobacter jejuni* diarrhea model in infant chickens. *Infect. Immun.* 43: 931–6.

Scherer, K., Bartelt, E., Sommerfield, C. and Hildebrandt, G. (2006). Quantification of *Campylobacter* on the surface and in the muscle of chicken legs at retail. *J. Food Prot.* 69: 757–61.

Sevolan, M. C., Winterfield, R. W. and Goldman, C. O. (1958). Avian infectious hepatitis; I. Clinical and pathological manifestations. *Avian Dis.* 2: 3.

Shane, S. M. (1992). The significance of *Campylobacter jejuni* infection in poultry: a review. *Avian Pathol.* 21: 189–213.

Shaughnessy, R. G, Meade, K. G, Cahalane, S., Allan, B., Reiman, C., Callanan, J. J. and O'Farrelly, C. (2009). Innate immune gene expression differentiates the early avian intestinal response between *Salmonella* and *Campylobacter*. *Vet. Immunol. Immunopathol.* 132: 191–8.

Singh, K. M., Shah, T., Deshpande, S., Jakhesara, S. J., Koringa, P. G., Rank, D. N. and Joshi, C. G. (2012). High through put 16s rRNA gene-based pyrosequencing analysis of the fecal microbiota of high FCR and low FCR broiler growers. *Mol. Biol. Rep.* 39: 10595–602.

Simaluiza, R. J., Toledo, Z., Ochoa, S. and Fernández, H. (2015). The prevalence and antimicrobial resistance of *Campylobacter jejuni* and *Campylobacter coli* in chicken livers used for human consumption in Ecuador. *J. Anim. Vet. Adv.* 14: 6–9.

Skirrow, M. B. and Blaser, M. J. (2002). Clinical aspects of *Campylobacter* infection. In *Campylobacter* 2nd Edition, Edited by I Nachamkin and M Blaser, ASM Press, Washington, DC.

Smith, C. K., Abuoun, M., Cawthraw, S. A., Humphrey, T. J., Rothwell, L., Kaiser, P., Barrow, P. A. and Jones, M. A. (2008). *Campylobacter* colonization of the chicken induces a proinflammatory response in mucosal tissues. *FEMS Immunol. Med. Microbiol.* 54: 114–21.

Stern, N. J., Cox, N. A., Bailey, J. S., Berrang, M. E. and Musgrove, M. T. (2001). Comparison of mucosal competitive exclusion and competitive exclusion treatment to reduce *Salmonella* and *Campylobacter* spp. colonization in broiler chickens. *Poult. Sci.* 80: 156–60.

Tudor, D. C. (1954). A liver degeneration of unknown origin in chickens. *J. Am. Vet. Med. Assoc.* 125: 219–20.

Van Deun, K., Haesebrouck, F., Heyndrickx, M., Favoreel, H., Dewulf, J., Ceelen, L., Dumez, L., Messens, W., Leleu, S., Van Immerseel, F., Ducatelle, R. and Pasmans, F. (2007). Virulence properties of *Campylobacter jejuni* isolates of poultry and human origin. *J. Med. Microbiol.* 56: 1284–9.

Van Deun, K., Pasmans, F., Ducatelle, R., Flahou, B., Vissenberg, K., Martel, A., Van den Broeck, W., Van Immerseel, F. and Haesebrouck, F. (2008). Colonization strategy of *Campylobacter jejuni* results in persistent infection of the chicken gut. *Vet. Microbiol.* 130: 285–97.

Veterinary Laboratories Agency (2005). VLA surveillance report. *Vet. Rec.* 157: 399–402.

Wagenaar, J. A., van Bergen, M. A., Newell, D. G., Grogono-Thomas, R. and Duim, B. (2001). Comparative study using amplified fragment length polymorphism fingerprinting, polymerase chain reaction genotyping and phenotyping to differentiate *Campylobacter fetus* strains from animals. *J. Clin. Microbiol.* 39: 2283–6.

Whenham, G. R., Carlson, H. C. and Aksel, A. (1961). Avian vibrionic hepatitis in Alberta. *Can. Vet. J.* 2: 3–7.

Whyte, R., Hudson, J. A. and Graham, C. (2006) *Campylobacter* in chicken livers and their destruction by pan frying. *Lett. Appl. Microbiol.* 43: 591–5.

Wigley, P. (2013). Immunity to bacterial infection in the chicken. *Dev. Comp. Immunol.* 41: 413–17.

Wigley, P. and Humphrey, S. (2014). The long road ahead: unravelling the immune response to *Campylobacter* in the chicken. In S. K. Sheppard (Ed.), *Campylobacter Ecology and Evolution*. Caister Academic Press, London.

Williams, L. K., Sait, L. C., Trantham, E. K., Cogan, T. A. and Humphrey, T. J. (2013). *Campylobacter* infection has different outcomes in fast and slow growing broiler chickens. *Avian Dis.* 57: 238–41.

Williams, L. K., Trantham, E. K. and Cogan, T. A. (2014). Conditional commensalism of *Campylobacter* in poultry. In S. K. Sheppard (Ed.), Campylobacter *Ecology and Evolution*. Caister Academic Press, London.

Wilson, D. L., Rathinam, V. A. K., Qi, W., Wick, L. M., Landgraf, J., Bell, J. A., Plovanich-Jones, A., Parrish, J., Finley, R. L., Mansfield, L. S. and Linz, J. E. (2010). Genetic diversity in *Campylobacter jejuni* is associated with differential colonization of broiler chickens and C57BL/6J IL10-deficient mice. *Microbiology* 156: 2046–48.

Withange, G. S., Wigley, P., Kaiser, P., Mastroeni, P., Brooks, H., Powers, C., Beal, R., Barrow, P., Maskell, D. and McConnell, I. (2005). Cytokine and chemokine responses associated with clearance of a primary *Salmonella enterica* serovar Typhimurium infection in the chicken and in protective immunity to rechallenge. *Infect. Immunity.* 73: 5173–82.

Zweifel, C., Zychowska, M. A. and Stephan, R. (2004). Prevalence and characteristics of Shiga toxin-producing *Escherichia coli, Salmonella* spp. and *Campylobacter* spp. isolated from slaughtered sheep. *Int. J. Food Microbiol.* 92: 45–53.

Chapter 2

Gastrointestinal diseases of poultry: causes and nutritional strategies for prevention and control

Raveendra R. Kulkarni, North Carolina State University, USA; Khaled Taha-Abdelaziz, University of Guelph, Canada and Beni-Suef University, Egypt; and Bahram Shojadoost, Jake Astill and Shayan Sharif, University of Guelph, Canada

1 Introduction

The growing global population and a need to meet the current as well as the future demand for high-value animal protein have put immense pressure on the livestock industry, including poultry, to enhance food animal production. According to a recent WHO/FAO joint report, annual meat production is projected to increase from 218 million tonnes in 1997-1999 to 376 million tonnes by 2030 (Salter, 2017). In this context, more efficient and sustainable food production systems that provide economically viable feed conversion efficiencies while also reducing the ecological footprint of livestock production are needed. Successful animal production depends largely on efficient feed conversion. Considering that the intestine is the primary site for digestion and nutrient absorption and that it is also the primary point of contact between the host and the external environment, including infectious agents, it is important to recognize that 'gut health' forms an integral part of a sustainable food animal production system. Therefore, maintaining a healthy intestinal environment is a critical element in ensuring the overall health and productivity of poultry flocks.

The prophylactic use of antibiotics as growth promoters (AGPs) has been a management practice for several decades in poultry production. However, the use of antibiotics in livestock may lead to the emergence of antibiotic resistance in bacteria. Therefore, restrictions on dietary antibiotics for poultry have been imposed in several countries, and

http://dx.doi.org/10.19103/AS.2019.0059.11

several other jurisdictions have plans to phase out AGPs in the near future. Restrictions on antibiotic usage may have negative consequences for performance, animal welfare, and general health of poultry, particularly for gastrointestinal (GI) disorders (Suresh et al., 2018). AGPs have been effective at reducing the burden of enteric infections, including necrotic enteritis (NE), in chickens. As a result, removal of AGPs to offer an 'antibiotic-reduced/free' environment in broiler production has posed a threat to the industry as it has led to a spike in enteric infections, which has significant negative implications for the industry. In this chapter, important enteric diseases or disorders will be discussed, briefly highlighting their etiology followed by possible nutritional interventions, including feed additives, as possible alternatives to AGPs for disease control.

2 Gastrointestinal (GI) tract diseases

Gastrointestinal (GI) diseases affecting poultry can be of infectious or noninfectious origin (Table 1). The etiological factors can range from infectious agents, environmental factors, and management practices, including feed and water, each of which can adversely affect the growth rate and feed conversion efficiency (Szkotnicki, 2013).

2.1 Infectious diseases

As the GI tract provides a large mucosal surface area allowing for digestion and absorption of ingested feed, it also poses a greater risk of being exposed to a variety of infectious agents. Enteric diseases, in many instances, are complex due to the involvement of more than one infectious agent, including bacterial, viral, fungal, or parasitic microorganisms (Weber et al., 2016). Infectious agents can gain access to poultry via different routes, such as oral or aerosol, and spread within or between farms can occur through contamination of feed, water, litter, fomites, or air. It is of note that the vertical route of transmission is also a major threat for introducing certain infectious agents, such as *Salmonella* or *Escherichia coli* during the early days of chick life (Calnek, 2015). Some of the important enteric disease agents that affect poultry are discussed briefly here with an emphasis on two economically important diseases: NE and coccidiosis.

2.1.1 Bacteria

Low-grade damage to the intestinal tract by pathogenic bacteria may cause poor feed conversion and a decreased rate of body weight gain in poultry flocks. More severe enteric damage by bacterial infections results in overt disease and high mortality (Adedokun and Olojede, 2018). Of the bacterial pathogens, *Salmonella* and Clostridia are considered the most important bacteria that affect poultry. To a certain extent, Campylobacteriosis caused by *Campylobacter jejuni*, *C. coli*, and *C. lardis* also poses intestinal health problems such as distension of the intestinal tract and diarrhea. However, most chickens carry *Campylobacter* asymptomatically and the importance of *Campylobacter* in poultry is due to its foodborne zoonotic potential (Marotta et al., 2015).

Table 1 Some of the important etiological agents/factors affecting GI health of poultry

| | | | | Noninfectious origin | | |
| | | | | Management | | |
Bacteria	Viruses	Parasites	Fungi	Environment	Feed	Water
Salmonella	Rota	Coccidia	Candida	Stocking density	Type	Quality
E. coli	Reo	Histomonas		Temperature	Quality	Palatability
Clostridia	Entero	Worms		Humidity, Air	Palatability	
	Astro			quality	Content	
	Corona			(ventilation and	Mycotoxins	
	Adeno			ammonia)		
	Influenza			Feeder/waterer		
	Paramyxo			placement		

(Infectious origin spans Bacteria, Viruses, Parasites, Fungi columns)

2.1.1.1 Salmonellosis

Salmonellosis in poultry, also sometimes referred to as pullorum or bacillary white diarrhea, is a serious disease of chickens of all ages caused by *Salmonella enterica* serovar Pullorum. While the disease is acute in young chicks, consisting of severe clinical signs and mortality/morbidity that peaks at 100% in 7–10-day-old birds, older birds exhibit chronic infection and remain subclinical (Wigley et al., 2005). The pathogen can be vertically spread through the egg (transovarian) or on the egg surface (by fecal contamination), or by feed and water contamination, or by bird-to-bird (horizontal) transmission (Berchieri et al., 2001, Barrow et al., 2012). It is noteworthy that pullorum has been categorized as a notifiable disease and has been eradicated from most commercial flocks around the world; however, disease incidence is still common in backyard and commercial flocks in some developing countries (Barrow et al., 2012). Certain serovars of *Salmonella* such as *Salmonella enterica* serovar Typhimurium and *Salmonella enterica* serovar Enteritidis are carried by poultry asymptomatically but pose a great zoonotic threat to humans.

2.1.1.2 Necrotic Enteritis (NE)

Clostridial infections are a major problem in poultry, and *C. perfringens*, the most important *Clostridium* species, can cause several clinical manifestations and lesions, including NE, necrotic dermatitis, ulcerative enteritis, and cholangiohepatitis, as well as gizzard erosion (Lovland and Kaldhusdal, 1999). Of these, NE has been considered the most economically important enteric disease in recent years due to the increasing demand for restrictions on the use of AGPs in poultry production. NE is a multifactorial disease and is of utmost importance to broiler production, as economic losses in the global poultry industry due to NE, both clinical and subclinical, are estimated to be $6 billion/year according to a recent report (Goossens et al., 2017). The clinical form of the disease often results in necrosis of the small intestine and is associated with high mortality. In addition to mortality, other economic losses due to NE, particularly in the case of subclinical NE, are attributed to reduced weight gain, higher feed conversion, and overall poor performance. These problems are due to chronic intestinal mucosal damage resulting in diminished nutrient digestion and absorption.

Clostridium perfringens bacteria are found widespread in breeder farms, hatcheries, grow-out houses, and processing plants, and are considered to be part of the normal intestinal flora of chickens (Craven et al., 2003). Some of the predisposing factors of NE include dietary contents, such as high animal protein and cereal grains (wheat, barley, and rye), that induce high viscosity of intestinal contents, in addition to other factors such as infection with coccidia or other mucosal pathogens. NE lesions can be among the most severe of any disease that affects the chicken intestine (Al-Sheikhly and Truscott, 1977). The pathogenesis of NE is complex because of the involvement of many microbial factors such as enzymes, adhesion molecules, house-keeping molecules, and importantly, tissue degrading toxins such as NetB, alpha-toxin, and TpeL, all of which contribute to virulence of *C. perfringens* (Goossens et al., 2017; Prescott et al., 2016). Recent reports also suggest that there are NE-causing strains that possess certain signature NE-associated virulence gene(s) that are absent in commensal avirulent non-NE-causing isolates of *C. perfringens* (Van Immerseel et al., 2016; Lepp et al., 2010). Intestinal mucosal damage that occurs during coccidiosis in chickens is usually considered as one of the most important predisposing factors, as coccidiosis often occurs just before or concurrently with outbreaks of NE in the field (Mot et al., 2014). However, in turkeys, mucosal damage is usually caused by coccidiosis, in addition to ascaridiasis, and viral hemorrhagic enteritis (HE) (Gazdzinski and Julian, 1992; Palya et al., 2007).

Although NE is very common in broilers, NE outbreaks can also occur in laying hens, particularly near the onset of laying or during the peak of production (Dhillon et al., 2004). Affected layers are weak, depressed, and have reduced egg production. Moreover, these birds show diarrhea with occasional high mortality with necrotic lesions. Cases in pullets have also been reported to be on the rise, particularly in pullets predisposed to coccidial infections (Hofacre et al., 2018). Importantly, an increasing demand for egg producers to adopt cage-free rearing systems is likely to challenge birds with more GI health issues including NE (Elwinger et al., 1992; Fossum et al., 1988). Antibiotics are the current best choice of treatment; however, many of them leave residues in eggs, forcing a withdrawal period that leads to economic losses. It is also often difficult to control NE with antibiotics as the disease progresses very rapidly sometimes producing irreversible intestinal damage. Hence, it is best to prevent NE rather than treat it.

2.1.2 Viruses

Most viral enteric infections occur in the first 3 weeks of life, which may cause diarrhea. These infections cause high morbidity but low mortality. Viral enteric infections may facilitate bacterial replication and attachment to the gut membrane, which can make the condition more severe. Determining the causative agent of enteritis or enteric disease is often complicated by the presence of more than one infectious agent, including combinations of bacteria, viruses, and parasites. For example, reoviruses have been isolated from flocks exhibiting enteric problems (Benavente and Martinez-Costas, 2007); however, whether this virus is the primary agent is questionable. In this context, it is clear that reoviruses can interact with other infectious agents of chickens such as *E. coli*, infectious bursal disease virus, and *Eimeria* resulting in increased pathological effects and economic losses (Weber et al., 2016; Benavente and Martinez-Costas, 2007). Reoviruses may be one of the several possible causative agents associated with various malabsorption syndromes, such as poult enteritis complex (PEC) and poult enteritis mortality syndrome (PEMS) in young turkeys, and runting-stunting syndrome (RSS) in broiler chickens (Mettifogo et al., 2014). Reovirus infections in broiler chicks result in viral arthritis, and a general lack of performance including diminished weight gain, poor feed conversions, chronic feed passage problems, uneven growth rate, and reduced marketability and sometimes mortality (Clavijo and Florez, 2018). Investigations of enteric diseases in chickens and turkeys have also focused on turkey coronavirus (TCoV), turkey and chicken astrovirus, avian orthoreovirus, avian rotavirus, torovirus, parvovirus, and several unknown 'small round viruses' (Dhama et al., 2015). Of note, a subset of these viruses has been isolated from birds affected by viral arthritis-tenosynovitis, stunting syndrome, respiratory disease, enteric disease, immunosuppression, and malabsorption syndrome (Cortez et al., 2017). Generally, virus-induced enteritis results in reduced daily weight gain, impaired feed conversion, and decreased flock uniformity (Guy, 1998).

Although enteric viral infections are commonly seen in young birds, older age groups can also be affected. For example, HE caused by type II adenovirus is an acute viral enteric disease of older turkeys and is characterized by depression, bloody droppings, and death (Beach et al., 2009). Importance of HE in turkeys is also related to the immunosuppressive nature of the virus which may exacerbate other diseases. However, the outcome of disease depends on many factors, including the age and immune status of affected birds, the virulence of the infecting virus, the presence of other infectious agents, nutrition, management practices, and environmental factors (Guy, 1998). Overall, diagnosing the causative agent of the malabsorption syndrome in poultry is challenging due to the possible involvement of more than one virus.

2.1.3 Parasites

Parasites that pose a serious challenge to poultry production are often internal parasites that include largely the protozoa. Commercial poultry are often infested with protozoan parasites, some of which can cause moderate-to-severe disease. Of utmost significance to the poultry industry, coccidiosis is one of the most common and economically important diseases of chickens worldwide, as described in detail in the next section. It is noteworthy that the intensive poultry production systems and high-density flocks increase bird susceptibility to protozoan parasites that have short and direct life cycles, such as coccidiosis (Williams, 2005). The other important protozoan disease that affects poultry is histomoniasis or blackhead, which is caused by *Histomonas meleagridis*. This pathogen causes severe lesions in the ceca and liver of many gallinaceous birds, of which turkeys are the most susceptible population (Abraham et al., 2014; McDougald et al., 2012; McDougald and Fuller, 2005). While histomoniasis in turkeys causes high mortality, sometimes approaching 100%, lower mortality rates of 10–20% are observed in chickens and many outbreaks may even go unnoticed. An interesting feature of this disease is that *H. meleagridis* is carried from host to host by eggs of the cecal worm *Heterakis gallinarum*. It is noteworthy that a previous observation indicated that the lesions of histomoniasis can be found more severe in turkeys when *C. perfringens* were present (Agunos et al., 2013).

Additionally, to a certain extent, helminths can also affect intestinal health in poultry, particularly those that are reared on used litter. Ascarids are the most common worms in poultry and *Ascaridia galli* is the most common ascarid, residing mostly in the intestinal lumen causing weight loss in birds (Jansson et al., 2010). *A. galli* has also been shown to transmit viruses and bacteria. Unlike cestodes and flatworms, the life cycle of Ascarids is direct. Economic losses due to Ascarid infections in chickens are largely attributed to parasite-induced anemia, retarded growth, and mortality.

2.1.3.1 Coccidiosis

Coccidiosis has been studied for decades and is a disease of the intestinal tract of almost all domestic and wild animals. For economic and disease control reasons, coccidiosis is one of the most important and challenging diseases of poultry, particularly in chickens (Chapman et al., 2013). The disease is caused by members of the *Eimeria* genus that are host-specific and do not require an intermediate host for their life cycle completion. Coccidia are tissue-trophic and multiply rapidly; their replication cycle involves different intermediate intracellular stages in different segments of the small and large intestine (Fatoba and Adeleke, 2018). These parasites undergo at least two asexual stages (schizogony) and one sexual stage (gametogony) during their replication cycle. The developmental stage of sporogony occurs outside of the host during which the oocysts undergo sporulation and the sporulated oocysts become infective to susceptible birds when they are ingested. The clinical disease results in severe lesions including erosive and hemorrhagic lesions in the intestinal segments leading to mortality, as well as impaired nutrient digestion and absorption, poor bird growth and performance, and mortality (Williams, 2005).

Many species of coccidia are widespread in countries where poultry are produced on a commercial basis. The spread of *Eimeria* from bird to bird and from flock to flock depends on the survival of oocysts of the parasite in the litter or soil. Chickens can be affected by nine species of *Eimeria*, of which six, namely *E. acervulina*, *E. maxima*, *E. brunetti*. *E. necatrix*, *E. mitis*, and *E. tenella*, are considered important. All avian *Eimeria* infect only one poultry species with the exception of *E. dispersa* which may infect and cause disease in turkeys, quails, and pheasants. Some of the coccidial members such as *E. maxima*, *E. necatrix* and *E. tenella* are deep tissue invaders and can cause severe necrosis, hemorrhage of the intestinal mucosa, and bloody diarrhea resulting in mortality (Barbour et al., 2015). Culled birds often appear pale due to anemia, and prior to death birds exhibit depression, poor weight gain and feed conversion, and a drop in egg production.

The presence of other microorganisms in chickens can impact coccidial infections, for example, it has been shown that certain indigenous bacterial species such as *Streptococcus faecalis*, *E. coli*, *Lactobacillus* species, and *Bacteroides* species may play a role in pathogenesis of cecal coccidiosis (Bradley and Radhakrishnan, 1973). It has also been shown that immunosuppressive diseases or conditions may act in concert with *Eimeria* to produce a more severe coccidiosis. Examples include Marek's disease and infectious bursal disease, both of which may interfere with the development of immunity to coccidiosis, and thus, may influence the severity of disease outcome (Williams, 2005). Coccidial infection in chickens is also thought to affect the outcome of other infectious diseases; for example, cecal coccidiosis may contribute to increased severity of histomoniasis in chickens (McDougald and Fuller, 2005). Additionally, in the context of NE, it is well known that subclinical coccidiosis acts as an important contributing factor in facilitating the growth and multiplication of *C. perfringens* and thereby, development of NE in broiler chickens by inducing mucosal damage (Al-Sheikhly and Truscott, 1977; Williams, 2005).

2.2 Noninfectious factors

2.2.1 Management and environment

Good management practices are key to successful poultry production. These practices include elements such as the design and structure of poultry barns, environmental conditions (ventilation, temperature, and litter condition), stocking density, feed and water supply, as well as the knowledge and experience of the people who manage the poultry operations (Powers and Capelari, 2017; Kiarie and Mills, 2019). These factors affect each other and can promote or adversely influence the health of the flock.

2.2.2 Toxins

Mycotoxins, the toxic secondary metabolites of fungi are one of the major causes of enteric disease in poultry (Guerre, 2016), and their presence in poultry feed has been identified as a widespread cause of economic losses due to impaired health status and reduced performance (Sklan et al., 2003). Although the effects of mycotoxins vary depending on the age and type of the bird, some common effects are reduced feed efficiency, growth performance, immunity, egg production, and hatchability along with increased mortality and organ lesions (mainly liver and kidney). There are over 100 known mycotoxins, and when the moisture content of grains rises, fungal growth and toxin production can result, leading to consumption of toxin-containing feed and subsequent disease (Greco et al., 2014). Mycotoxins include aflatoxin produced by *Aspergillus* sp., trichothecene toxins (T-2), diacetoxyscirpenol (DAS), and deoxynivalenol (DON or Vomitoxin) produced by *Fusarium* sp., and ochratoxin produced by *Aspergillus* sp. While birds with aflatoxicosis typically demonstrate hemorrhages in the intestinal tract, muscles, and skin along with enlarged kidneys and liver, the effects of trichothecenes include hemorrhagic and necrotic lesions in the gizzard and proventriculus, and also atrophy of the bursa of Fabricius and thymus (Barnes et al., 2001). In combination with other factors, mycotoxins can predispose or exacerbate outbreaks of enteric diseases. For example, it has been shown that ochratoxin A and coccidial infections can interact to adversely affect broiler growth and performance (Koynarski et al., 2007).

2.2.3 Diet and nutrition

Not only is the GI tract responsible for physiological functions including digestion and absorption of macro- and micronutrients, it also acts as a physical barrier between the host and the environment, underscoring the importance of GI health to protect against invading pathogens (Kiarie and Mills, 2019) and nutrition can influence poultry GI health and bird susceptibility to enteric diseases. Factors associated with diet, such as feed palatability, quality, and type of feed as well as feeding strategies including restricted feeding, can negatively affect GI health by altering intestinal microbiota and/or enzymatic activity leading to enteric disorders (Gelli et al., 2017). Additionally, the amount of fiber and non-starch polysaccharides (NSP) in the feed, and the content and quality of the ingredients used in the feed, are also key factors in maintaining intestinal integrity. Dietary fibers from plant cell walls, NSP, and lignins have the capacity to absorb water in the intestine leading to higher amounts of bulky material in the intestine, thus increasing the viscosity and decreasing the digestibility of food (Bach Knudsen et al., 2017). For example, cereals used in poultry diets contain various levels of NSP such as β-glucans and arabinoxylans that resist digestion and increase the viscosity of the lumen contents, thus increasing the digesta retention time and facilitating bacterial colonization in the small intestine (Bach Knudsen et al., 2017). Diets that include barley, wheat, rye, and oat contain high amounts of NSP, which has been associated with an increased number of *C. perfringens* in the chicken intestine and higher mortality due to NE (Annett et al., 2002). To this end, it is of note that that in commercial poultry diets, where these cereals are used, NSP-degrading enzymes such as xylanase, beta-glucanase, mannanase, and galactosidase are used to decrease viscosity and eliminate the negative effects of NSP (Choct et al., 1995; Ferrandis Vila et al., 2018).

Dietary protein content is also an important factor that influences GI health in birds. Different protein sources of animal and plant origin are used to meet the protein demands of poultry, and some protein sources could play an important role in predisposing chickens to certain intestinal diseases. For example, chickens fed with diets containing high amounts of animal protein have been shown to have a higher number of *C. perfringens* in the intestine (Drew et al., 2004), and a higher amount of fish meal in feed may predispose broilers to NE (Shojadoost et al., 2012). Additionally, the physical texture of the feed can influence digestion and thereby affect bird susceptibility to enteric diseases. For example, broiler chickens fed with finely ground wheat showed elevated mortality due to NE and coccidiosis compared to chickens fed a coarsely ground wheat-based diet (Choct et al., 1996). Subsequent reports also suggested that feeding whole wheat to broiler chickens can reduce the intestinal burden of *Salmonella* and *C. perfringens* (Engberg et al., 2004). To this end, Korver et al. (2004) suggested that when the GI tract is in a healthy state, inclusion of whole wheat into the diet may help improve digestive tract function, but when the integrity of the intestinal tract is impaired, inclusion of whole wheat may decrease performance of the GI tract. Apart from the texture, factors such as excessive dust levels in the feed can reduce its palatability and improperly stored feed may allow fungal growth and rancidity, thus affecting its nutritive content and leading to impairment of GI health (Kiarie and Mills, 2019). Additionally, inadequate feeder or water space and distribution can result in a non-uniform feed/water intake and, thus, lead to precipitation of enteric disease (Powers and Capelari, 2017).

3 Nutritional interventions

Over the last decade or so, much research has focused on devising dietary strategies for controlling GI diseases or disorders in poultry. These include either manipulation of feed composition (ingredients and nutritional content) or finding feed additives to replace AGPs (Adedokun and Olojede, 2018; Morrissey et al., 2014). In this section, we have discussed both nutritional modulation and research on feed additives focused on enhancing GI health and resistance to enteric infections in poultry. Some of the nutritional components of poultry feed that have been extensively studied in the context of GI health issues are dietary fiber, starch, protein, and fat. Important effects of these constituents on GI health and disease are discussed.

3.1 Dietary modulation

3.1.1 Dietary fiber

Dietary fiber content of poultry feed and its effects on GI health and the resident microflora have been studied in detail (Leeson et al., 1997; Morrissey et al., 2014). Studies focused on water-insoluble fiber indicate positive effects on nutrient digestion despite the fact that insoluble fiber has only a limited nutritional value in poultry due to its low fermentability (Carre et al., 1995). A previous study showed that dietary supplementation with 10% oat hulls can significantly increase gizzard weight and pancreatic enzyme activity (Gonzalez-Alvarado et al., 2008). Furthermore, these authors also showed that the oat hulls can help overcome the adverse effects of poor intestinal conditions caused by soluble NSP by reducing digesta moisture content. It is noteworthy that the high-molecular-weight soluble NSP, which increase the viscosity of luminal content, can adversely affect GI tract development and nutrient digestion and thus can have a negative impact on production performance and intestinal health (Langhout et al., 2000; Maisonnier et al., 2003). As mentioned earlier, increased mucus production might exacerbate overgrowth of mucolytic bacteria such as *C. perfringens* in the small intestine; therefore, the mucus-increasing fiber components, such as NSP, and anti-nutritional factors, such as phytate, need to be taken into consideration when formulating feed (Paiva et al., 2013). In this context, it has been proposed that low levels of NSP along with optimal insoluble fiber levels may help modify the intestinal microbiota through mechanisms such as production of volatile butyrate fatty acids and thus, resist *C. perfringens* and coccidial infections (Zubair et al., 1996).

3.1.2 Dietary protein

With the growing trend of phasing out AGPs, dietary management of GI health has become critical in poultry production, and the inclusion of high-quality protein in poultry diets is an important dietary management strategy. From a nutritional standpoint, animal by-products (fish meal, meat, and bonemeal) are generally considered of much higher quality than plant protein sources because the amino acid balance of the animal products more closely reflects the nutritional requirements of the bird (Beski et al., 2015). However, it is also of note that overheating of animal by-products during the rendering process may reduce the digestibility/availability of amino acids. Low-quality protein in the diet is associated with poor protein digestibility and may result in the accumulation of large amounts of undigested protein in the hindgut, stimulating the growth of proteolytic and potentially pathogenic bacteria including *C. perfringens*. In this context, several studies have examined the relative effects of animal proteins (fish meal, meat, and bonemeal and feather meal) or vegetable proteins (potato, pea and soy protein concentrates, or corn gluten meal) on *C. perfringens* counts in the ileum and cecum (Dahiya et al., 2007; Palliyeguru et al., 2010). The conclusions from these studies indicated that vegetable protein sources, except for potato protein concentrate, could reduce *C. perfringens* counts and that the dietary glycine was very critical such that *C.*

perfringens numbers were positively correlated with glycine content in the diet and in the intestinal contents (Dahiya et al., 2005). To this end, while the crude protein levels in the diet may be reduced to offer a better intestinal health, it is equally important to incorporate the required levels of important amino acids, such as glycine. Therefore, the use of feed ingredients that can reduce incidence of intestinal disorders would be helpful for diet optimization without in-feed antibiotics.

3.1.3 Dietary fats

Dietary fats and fatty acids are also important nutritional factors that affect GI health and their effects are highly dependent on quantity, type, and quality of fat and fatty acid composition. Studies have indicated that unsaturated lipids are better digested than saturated lipids and that certain components such as soya, coconut oils, and milk lipids tend to reduce the *C. perfringens*-induced NE in broiler chickens (Gilbert et al., 2018). A previous report showed that fat digestibility during coccidial and *C. perfringens* infections is generally reduced and that the use of coconut oil in diets could ameliorate the disease due to its superior digestion compared to other oils (Moore, 2016; Adams et al., 1996). Persia et al. (2006) observed that 15% fish meal in a broiler diet can ameliorate *E. acervulina* infection and increase amino acid digestion and absorption; this was likely related to the effect fish oil has on ameliorating infection-induced inflammatory responses (Persia et al., 2006). Therefore, it seems that impaired intestinal health requires a change in energy source from saturated fats to sources containing unsaturated fatty acids or medium-chain fatty acids and that unsaturated vegetable oils or oils that contain high levels of medium-chain fatty acids have antimicrobial effects.

3.2 Feed additives

Antibiotics have been used in the poultry industry for decades to prevent disease and to improve feed efficiency and growth. However, concerns about increasing antimicrobial resistance among bacteria and the threat of its spread through the food chain into the human population demands the need for effective alternatives to antibiotics (Mehde et al., 2018). Among a series of experimental alternatives, some have shown to be effective in enhancing GI health and helping with resistance against enteric infections, including plant-derived extracts (essential oils (EO)), prebiotics, probiotics, and organic acids (OA), and they are described and discussed in the next section.

3.2.1 Essential oils (EO)

Essential oils (EO) that constitute a major class of plant-derived extracts have been shown to offer beneficial effects for poultry health, immunity, and feed digestion (Farhadi et al., 2017; Hernandez et al., 2006). EO, which are secondary metabolites of plants, and can be extracted from different plant parts, are known to possess antimicrobial properties and have been suggested as potential replacements for antibiotics for use in livestock (Burt, 2004; Varel et al., 2007). Additionally, they are shown to exert antioxidant activity and, thus, help in boosting host immune functions (Kim et al., 2016; Shahid et al., 2018). Specific evidence in the context of EO-mediated antimicrobial activity for use as a feed additive in animal feeds also comes from studies showing antiviral (Garozzo et al.,

2009), antifungal (Sakkas and Papadopoulou, 2017), and antiparasitic (Pessoa et al., 2002) activities. Furthermore, due to their aromatic properties, they have been shown to have a positive impact on reducing odors and ammonia levels in poultry houses, thus improving bird performance (Varel et al., 2007). Of around 3000 known EO, around 300 are available on the market, of which many are available for use in the livestock industry including poultry (Kalemba and Kunicka, 2003). The main components of EO are terpenoids (citronellol and menthol), aliphatic hydrocarbons (phenols and thymols), and aromatic organic compounds (cinnamaldehyde and phelandral) (Delaquis et al., 2002). The activity of EO has been mainly attributed to the major components that comprise 85% of the herbal extract, of which phenolic compounds are the key component that offers antibacterial activity (Cosentino et al., 1999). In addition to enhancing feed palatability, EO are able to assist in digestion through the induction of GI peristaltic movements and secretion of saliva and digestive enzymes (Zhai et al., 2018).

3.2.1.1 Essential oils and antimicrobial activity

Effects of EO on various bacterial, viral, and fungal pathogens have been widely reported (Burt, 2004; Kalemba and Kunicka, 2003; Mith et al., 2014; Sakkas and Papadopoulou, 2017). The EO extracted from bay, cinnamon, clove, and thyme were shown to reduce the growth of bacteria *in vitro* including those foodborne pathogens, namely *C. jejuni*, *Salmonella enterica* serovar Enteritidis, and *E. coli* (Smith-Palmer et al., 1998). Another study that used the non-water-soluble subfraction of EO (carvacrol and thymol) showed significant *in vitro* antimicrobial activity against *E. Coli*, *Enterococcus faecalis*, *Pseudomonas aeruginosa*, *Salmonella enterica* serovar Enteritidis, and *Streptococcus pyogenes* (Gulluce et al., 2003). Furthermore, there have been many reports that describe the effectiveness of EO against *C. perfringens*, the causative agent of NE in chickens. For example, EO combinations of thymol, eugenol, carvacrol, curcumin, and piperine showed a reduction in the number and intestinal colonization of *C. perfringens* in broiler chickens (Mitsch et al., 2004). Additionally, this study also showed that a combination of ginger oil and carvacrol EO can reduce intestinal lesions and improve broiler performance. In this context, the authors suggested that in addition to their direct effects on *C. perfringens*, EO could enhance the growth and colonization of intestinal microbiota that may inhibit proliferation of *C. perfringens*.

   In the context of foodborne pathogens present in the poultry gut, there has been ample evidence to suggest that EO can reduce the intestinal burden of *E. coli* and *Salmonella*. To this end, a recent study showed a reduction in the number of *Salmonella*, *E.coli*, and Clostridia in the cecum of broiler chickens, when a mixture of plant extracts (fennel, melissa balm, peppermint, anise, oak, clove, and thyme) was used as a feed supplement (Wati et al., 2015). It has also been shown that supplementation of broiler chicken feed with cinnamon oil reduced the number of *E. coli* in the prececal digesta without altering the number of lactobacilli (Gomathi et al., 2018). Furthermore, an EO blend (carvacrol, thymol, eucalyptol, and lemon) administered via drinking water was shown to reduce *Salmonella enterica* serovar Heidelberg counts in the crop (Alali et al., 2013). In addition to their effects on foodborne enteric bacterial pathogens, EO have also been shown to exert anti-protozoan activity in poultry, particularly against coccidia. Previous studies have found that EO can modulate intestinal microbiota populations in chickens that were challenged with a combination of *Eimeria* oocysts containing *E. acervulina, E. maxima, and E. tenella* (Hume et al., 2006; Oviedo-Rondon et al., 2006). Supplementation of poultry feed with eucalyptus and peppermint demonstrated a reduction in lesion scores and number of intestinal oocysts as well as decreased weight loss in birds challenged with eight prevalent *Eimeria* spp (*E. acervulina, E. brunetti, E. hagani, E. maxima, E. mivati, E. necatrix, E. praecox, and E. tenella*) (Barbour et al., 2015). The authors in this study concluded that this blend of EO could be used as an alternative to anticoccidial drugs to control coccidial infections in chickens. In a recent comprehensive *in vitro* study, Jitviriyanon et al. (2016) demonstrated an *in vitro* oocysticidal effect of ten EO extracted from different indigenous plants, of which some, especially those extracted from *Boesenbergia pandurata* and *Ocimum basilicum*, were able to induce degenerative changes in oocysts of *E. tenella* and inhibited sporulation (Jitviriyanon et al., 2016). Another study, using the herb extract from *Aloe secundiflora* leaves, also showed a marked reduction in clinical signs and lesions as well as a dose-dependent reduction in fecal oocyst counts in *E. tenella*-challenged chickens (Kaingu et al., 2017).

In summary, different EO have been shown to have antimicrobial activities against many bacterial and protozoan pathogens, including *C. perfringens, Salmonella, E. coli, C. jejuni*, coccidia, and others in the intestine of poultry, while also not affecting the normal microbiota population. Indeed, the use of EO seems promising for the poultry industry as a reasonable alternative for AGP usage. However, attention also needs to be paid to the fact that certain bacteria such as *Salmonella* Typhimurium, *Salmonella* Enteritidis, *E. coli, Staphylococcus aureus*, and *E. faecalis* can gain adaptation ability to resist antimicrobial activities of EO (oregano, cinnamon, and other oils), when used as feed additives (Becerril et al., 2012; Melo et al., 2015).

3.2.2 Prebiotics

Prebiotics are nondigestible oligosaccharide carbohydrate compounds, and examples include fructooligosaccharides (FOS, derived from grains), galactooligosaccharides (GOS, derived from milk), and mannan-oligosaccharides (MOS, derived from the cell wall of the yeast *Saccharomyces cerevisiae*), each of which have potential to improve poultry health and reduce the burden of enteric pathogens (Hughes et al., 2017; Ricke, 2018). Prebiotics exert their beneficial effects on the host through improvement of intestinal function, modulation of host immune responses, and modification of the gut microbiota (M'Sadeq et al., 2015a). In the context of their immunomodulatory activities, *in ovo* administration or dietary inclusion of GOS and inulin has been shown to reduce the expression of proinflammatory cytokines and signaling molecules in lymphoid organs of chickens (Hughes et al., 2017; Smirnov et al., 2005; Slawinska et al., 2016). Emerging data also indicate that dietary supplementation of oligosaccharides can reduce colonization of *C. perfringens, E. coli*, and *Salmonella* by selectively promoting proliferation of beneficial bacteria such as bifidobacters and lactic acid-producing bacteria and/or by blocking the sites of bacterial attachment on the intestinal epithelium (Kim et al., 2011; Pourabedin and Zhao, 2015; Teng and Kim, 2018). Furthermore, fermentation of these oligosaccharides by resident microbiota produces short-chain fatty acids, mainly acetate, propionate, butyrate, and other by-products, which, in turn, can enhance host defense against infections (Pourabedin and Zhao, 2015; Teng and Kim, 2018). Prebiotics have also been shown to be effective against enteric protozoan parasites. For example, supplementation of a *Saccharomyces cerevisiae* fermentation product and galactoglucomannan oligosaccharide-arabinoxylan provided significant protection against *E. maxima* and *E. acervulina* infections in broiler chickens (Faber et al., 2012; Lensing et al., 2012). However, the effectiveness of prebiotics in controlling NE in chickens has been somewhat less conclusive. While some studies have shown that dietary inclusion of arabinoxylo-oligosaccharides, MOS, or yeast cell wall extract can reduce NE lesions and mortality (Keerqin et al., 2017; M'Sadeq et al., 2015b), one study did not observe any beneficial effects when the same or other oligosaccharides were used (Hofacre et al., 2018). The reason for these discrepant findings could be due to the different types and dosage regimens of oligosaccharides and/or the strain *C. perfringens* used in these studies. Nonetheless, prebiotics seem to offer some beneficial effects as feed additives in reducing enteric pathogen burdens in poultry.

3.2.3 Probiotics

Probiotics are beneficial microbes that confer various health benefits to the host (Taha-Abdelaziz et al., 2018). Several species of the bacterial genera, such as *Lactobacillus*, *Streptococcus*, *Enterococcus*, *Enterobacter*, and *Bifidobacteria*, and yeasts, such as *Saccharomyces*, *Torulopsis*, *Aspergillus*, and *Candida*, have been widely used as beneficial microbes. In the context of poultry, in addition to their roles in improving bird performance and GI health, supplementation with probiotics or their by-products have been shown to provide protection against various foodborne and enteric pathogens such as *E. coli*, *Salmonella enterica* serovar Enteritidis, *C. perfringens*, *C. jejuni*, and coccidia (Nakphaichit et al., 2011; Pan and Yu, 2014; Pascual et al., 1999; Shin et al., 2008; Strompfova et al., 2010).

Accumulating evidence also suggests that compared to single-strain probiotics, multi-strain probiotics exhibit greater efficacy for improving intestinal and immune functions and controlling bacterial pathogens, such as *Salmonella*, and protozoan parasites, such as *E. tenella* (Chen et al., 2012; Lensing et al., 2012). This is attributed to synergistic interactions between probiotic strains in mixtures (Chapman et al., 2013). Additionally, routes of administration of probiotics such as in feed, water, spray, and also experimentally via oral gavage seem to influence their beneficial effects. For example, administration of probiotics in drinking water can improve bird performance and enhance resistance against mixed *Eimeria* infection (*E. acervulina*, *E. maxima*, and *E. tenella*) in broiler chickens compared to in-feed supplementation (Ritzi et al., 2014). This observation may be attributed to the enhanced viability of probiotics in water during passage through the GI tract as it shortens the gastric transit time, thereby reducing the negative impacts of gastric acid and digestive secretions.

Probiotics can exert their protective effects against microbial pathogens either directly by inhibiting their growth, attenuating their virulence, and by competing with them for space and nutrients (referred to as competitive exclusion), or indirectly by modulating the host immune system and intestinal microbiome composition (Adedokun and Olojede, 2018; Brisbin et al., 2010). Some of the mechanisms of probiotics in the context of certain important enteric pathogens of poultry are briefly discussed in the following section.

3.2.3.1 Competitive exclusion and inhibition of pathogen growth

Mounting evidence indicates that mucosal adhesion is an important prerequisite for probiotics to establish colonization and is regarded as a key element for selection of probiotic candidates (Tuomola et al., 2001). In light of the fact that competitive coexistence is one of the mechanisms of pathogen exclusion, a recent study has demonstrated a relative dominance of *Lactobacillus* and *Bifidobacterium* sp. in the intestinal microbiota of chickens when they were given as dietary supplements. The competition capacity of probiotic bacteria, however, may vary according to the probiotic strain and the microbial competitor. For instance, oral administration of two doses of 10^9 colony forming units of *Lactobacillus johnsonii* strain FI9785 was not sufficient to confer protection against *C. jejuni* (Manes-Lazaro et al., 2017), while a single dose of this strain was sufficient to attain a competitive advantage over *C. perfringens* (La Ragione et al., 2004). Another study has also shown that administration of a lower dose of *L. johnsonii* R-17504 reduces *Salmonella enterica* serovar Enteritidis colonization (Van Coillie et al., 2007). Nevertheless, it is noteworthy that co-administration of other products such as prebiotics, OA, and EO with probiotics has been shown to enhance their protective efficacy against a wide range of enteric pathogens, such as *C. perfringens* and *Eimeria* spp (Pourabedin and Zhao, 2015; Alagawany et al., 2018; Giannenas et al., 2012). The addition of these agents likely facilitates growth and proliferation of probiotic bacteria in addition to inducing immunomodulatory effects on the host immune system (Caly et al., 2017; Namkung et al., 2004).

Numerous studies have also demonstrated that probiotics and their metabolites display broad-spectrum bactericidal activity against a wide range of Gram-negative and Gram-positive bacteria when tested *in vitro* (Guo et al., 2017). For example, Teo and Tan (2005) demonstrated that a single strain of *Bacillus subtilis* exhibits broad-spectrum inhibitory activity against various strains of *C. perfringens*, *C. jejuni*, and *Campylobacter coli*. To this end, attempts have been made to identify the antimicrobial compounds produced by probiotic bacteria. Among these compounds, bacteriocins have been investigated and several bacteriocins have been shown to possess antagonistic activity against various strains of *C. perfringens*, including pediocin A, divercin, nisin, subtilin, and bacteriocin-like inhibitory substance (Caly et al., 2017; Dabard et al., 2001; Jozefiak et al., 2012; Sharma et al., 2014; Udompijitkul et al., 2012). The production of these active metabolites is thought to provide probiotics with a competitive growth advantage over pathogenic microorganisms.

3.2.3.2 Attenuation of virulence factors

Existing evidence indicates that motility and adhesion are essential factors for microbial colonization of the GI tract, and the alteration of the expression of genes responsible for these attributes could lead to a reduction in the ability of the pathogen to adhere to and colonize mucosal surfaces (Haiko and Westerlund-Wikstrom, 2013; Ribet and Cossart, 2015). In addition to effects on motility and adhesion, *in vitro* studies have shown that probiotic exposure of pathogenic bacteria, such as *C. perfringens*, *C. jejuni*, and pathogenic *Salmonella*, results in downregulation of genes responsible for invasion, biofilm formation, and toxin and auto-inducer production (Guo et al., 2017; Li et al., 2011; Muyyarikkandy and Amalaradjou, 2017; Najarian et al., 2019).

3.2.3.3 Improvement of intestinal morphology

Maintenance of intestinal mucosal barrier integrity is essential for the prevention of enteric infections. In addition to their role in the induction of innate responses in epithelial cells, probiotic supplementation has been shown to improve intestinal morphology and mucus barrier function (Aliakbarpour et al., 2012; Smirnov et al., 2005), thereby enhancing bird performance as well as providing resistance against enteric pathogens, including *C. perfringens* and coccidia. For example, when used without anticoccidials, different probiotic bacteria (*Enterococcus faecium*, *Bifidobacterium*, *L reuteri*, *L salivarius*, and *Bacillus subtilis*) have been shown to mitigate the negative effects of *Eimeria* in broiler chickens (Giannenas et al., 2012, 2014). Furthermore, combination of a probiotic mixture (*Enterococcus*, *Bifidobacterium*, *Pediococcus*, and *Lactobacillus*) with an anticoccidial vaccine was shown to provide enhanced protection against *Eimeria* infection in broilers compared to those that received vaccine-only controls (Ritzi et al., 2016; Lin et al., 2017). A recent study has also shown that *B. licheniformis* supplementation was not only able to restore intestinal integrity but also the ileum and cecal microbial balance in chickens challenged with *C. perfringens* (Lin et al., 2017).

3.2.3.4 Modulation of intestinal immune responses

In the context of mucosal immunity, it has been shown that lactobacillus probiotic supplementation enhances antigen-specific and natural antibodies and also alters the expression of cytokines, antimicrobial peptides, and T cell surface markers in gut-associated lymphoid tissue (GALT), which, in turn, may enhance resistance to bacterial and parasitic pathogens in chickens (Akbari et al., 2008; Brisbin et al., 2010; Haghighi et al., 2008). Administration of a mixture of probiotic bacteria to chickens resulted in a significant increase in antibody-mediated immune responses to antigens, such as sheep red blood cells (Haghighi et al., 2005). Further, Haghighi et al. (2006) observed an enhanced production of natural antibodies in probiotic-treated chickens (Haghighi et al., 2006). Additionally, oral administration of *Lactobacillus* in chickens has been shown to affect antibody- and cell-mediated immune responses (Brisbin et al., 2011, 2012). Some of these immune-enhancing effects may be attributed to the structural constituents of probiotic bacteria, including their DNA and cell wall components (Brisbin et al., 2008). It is possible that interactions between bacterial components, such as DNA and peptidoglycan, and cells of the immune system, various innate or adaptive immune pathways are triggered. For example, administration of probiotics to chickens results in alteration of cytokine and antimicrobial peptide gene expression in cecal tonsils after infection with *Salmonella* (Akbari et al., 2008). The *in vitro* effects of *Lactobacillus* species on chicken immune system cells have been demonstrated, marked by regulation of gene expression in subsets of cecal tonsil cells (Brisbin et al., 2011, 2012) and splenocytes (Brisbin et al., 2010), in addition to enhancing the function of macrophages (Brisbin et al., 2015). In the context of *Eimeria* infection, probiotic administration has been shown to increase the levels of T helper 1-type cytokines (interferon-gamma) in both serum and intestinal secretions and the number of IELs expressing certain cell surface markers (CD3, CD4, CD8, and αβ T cell receptor). These immune-enhancing activities are associated with attenuation of the virulence and reproductive capacity of *Eimeria* spp (Dalloul et al., 2003). Similarly, dietary supplementation of *Lactobacillus johnsonii* elicits intestinal mucosal immunity in the ileum and cecal tonsils associated with enhanced resistance against subclinical NE in broiler chickens (Wang et al., 2017). Although there is evidence that individual or combinations of probiotic bacteria can modulate immune system gene expression and, thus, the immune functions, it is still not very clear as to whether probiotics can completely replace AGP in commercial practice and that they can effectively mitigate severe infectious challenges.

Taken together, it is evident that probiotics may serve as an important component of strategies aimed at developing alternatives to AGP in poultry. Although a vast amount of research has been devoted to understanding the dynamic microbial ecosystem in the intestine and the host-microbe or microbe-microbe interactions, there is still considerable work required to dissect out the precise mechanisms by which probiotics can provide general or GI health benefits in poultry.

3.2.4 Organic acids (OA)

Organic acids (OA) belong to a broad class of compounds that have important roles in various fundamental metabolic processes of host physiological machinery. These compounds, including formic, fumaric, propionic, citric, and lactic acid, and their salts (e.g. calcium formate, calcium propionate), have been traditionally used in animal feeds for reducing bacterial and fungal growth (Dittoe et al., 2018). Considering their safety and antimicrobial activity, OA are classified as 'feed preservatives' in Europe (Adil et al., 2010). As their use in animal production has proven to be beneficial, increasing evidence accumulated over the last two decades has indicated that the use of OA could also contribute to increased weight gain, higher feed conversion rates, and reduced incidence of GI-related health issues in livestock (Mikkelsen et al., 2009). In poultry, OA have been used either in feed or in drinking water with the objective of reducing intestinal pathogen burden and the associated toxic microbial metabolites. This practice has also been shown to improve nutrient digestibility, thereby enhancing bird performance and immune health of the avian intestine (Diarra and Malouin, 2014). The antimicrobial effects of OA, including short-chain and medium-chain fatty acids, also seem to depend on both the concentration of the acid and the microbial pathogen exposed to the acid (Adil et al., 2010). These acidifiers are used to benefit production in three ways in poultry operations: (1) OA added to the feed facilitates prevention of bacterial or mold growth in feed and also reduces the pH in the crop, (2) OA given via drinking water inhibits microbial growth in the water and reduces the pH of the crop and intestinal contents to facilitate pathogen control, and (3) OA sprayed onto the poultry litter can affect the bacteria that facilitate the breakdown of uric acid and, thus, limit the amount of ammonia release (Rodjan et al., 2018).

OA treatments composed of individual acids and blends of several acids have been found to exert antimicrobial activities (Gadde et al., 2017). Important enteric bacterial pathogens that affect poultry including *Salmonella*, *Campylobacter*, *C. perfringens*, and *E. coli* have been shown to be controlled by supplementation of OA in feed or in water (Van Immerseel et al., 2006). For example, a study by Koyuncu et al. (2013) showed that treatment of pelleted and compound mash feeds with formic acid and different blends of formic acid, propionic acid, and sodium formate could significantly reduce *Salmonella* counts in the feed (Koyuncu et al., 2013). The current practice of drinking water acidification in the broiler industry has improved bird performance as well as reduced pathogen load in the water and in the crop and proventriculus, coupled with an optimal regulation of intestinal microflora and adequate digestion of feed (Dittoe et al., 2018). In support of this, Bourassa et al. (2018) showed that incorporation of OA (lactic acid, acetic acid, or formic acid) in the drinking water during pre-transport feed withdrawal can reduce *Salmonella, E. coli*, and *Campylobacter* contamination of crops and broiler carcasses at processing (Bourassa et al., 2018).

Similarly, mixtures of OA (fumaric acid, calcium format, calcium propionate, potassium sorbate, calcium butyrate, calcium lactate, and hydrogenated vegetable oil) were found to be more efficacious than an AGP (Enramycin) at decreasing intestinal *C. perfringens, E. coli*, and *Salmonella* spp. (Manafi et al., 2019). Fernandez-Rubio et al. (2009) also suggested that supplementation of 0.2% encapsulated OA to the diet might improve the proliferation of useful commensal microbiota (*Lactobacillus* spp.) and diminish the population of pathogenic bacteria in poultry intestinal contents (Fernandez-Rubio et al., 2009).

Collectively, the aforementioned evidence shows that OA supplementation of poultry feed or water can help reduce the load of enteric pathogens, including *E. coli, C. perfringens, Salmonella*, and *Campylobacter* and, thus, improve the health and performance of poultry. However, to what extent OA can effectively replace AGP in commercial production remains to be yet further investigated.

4 Conclusion and future trends

Careful and well-thought-out nutritional intervention strategies involving the use of feed ingredients and feed additives may be exploited to promote gut health and development, and to reduce GI disease burden in poultry. These strategies may exert their effects via three different pathways: (1) Enhancement of intestinal integrity and functions, including establishment of beneficial microbial population, (2) reduction of enteric pathogen burden, and (3) modulation of host immune responses. Clearly, strategies aimed at reducing the incidence of enteric infections are critical for the productivity, sustainability, and profitability of the poultry industry. Vast research in the last two decades has put forth several candidate feed additives that have potential as alternatives to AGP in poultry production. These include prebiotics, probiotics, phytogenic feed additives (EO), antimicrobial peptides, bacteriophages, antibodies, enzymes, and acids, each of which can impact the incidence and severity of GI diseases, including NE and coccidiosis. Although many approaches have been proposed, combination of more than one approach seems necessary to enhance the performance and GI health of poultry. Some examples include choosing feed ingredients that have higher digestibility and nutritional value, a combination of prebiotics, probiotics, and EO that are known to benefit gut health and, importantly, proper management practices. However, a challenging question that still remains to be answered is, whether these AGP replacements can actually protect the birds in the face of a serious challenge, particularly in a commercial poultry setting? Many studies have shown the efficacy of AGP alternatives under a low-challenge environment and that such testing under experimental conditions in the absence of an active infectious challenge somewhat limits their potential applications. Furthermore, these AGP alternative strategies or approaches will require suitable adaptation to existing feeding programs and poultry production practices considering the wide variation in global climate and in housing and management practices. Nevertheless, given the increasing global demand for high-quality protein for humans from poultry sources, combined with the looming potential for decreases in production as AGP are removed from the industry, employing new nutritional intervention strategies is essential to enhance the GI and overall health of farmed poultry while also reducing the impact of enteric diseases to a considerable degree.

5 Where to look for further information

An introduction to the topic of GI diseases as well as a thorough understanding of the disease pathophysiology and microbiology aspects can be found in the book titled 'Diseases of Poultry', 14th edition by David E Swayne (Editor-in-chief) published by Wiley Blackwell publishers. Considerable insight and detailed information about the role of nutrition and enteric microbiota in the maintenance of optimal gut-health and the future roadmap for the use of various types and forms of feed additives, including probiotics, as antimicrobial alternatives can be obtained from two review articles authored by Adedokun and Olojede (2018) and Clavijo and Florez (2018), which are cited here in this chapter.

Additionally, the United States Department of Agriculture (USDA), which is the largest funding agency for animal researchers, provides up-to-date information about various topical aspects of animal health and disease surveillance and prevention, including those affecting poultry (https://www.usda.gov/topics/animals). In particular, the USDA's National Institute of Food and Agriculture (NIFA) that provides ample information highlighting important animal research areas, including the area of 'reduced use of antibiotics' has been an encouraging source for many poultry researchers to develop high impact research programs (https://www.nifa.usda.gov/topic/animal-health). Furthermore, an online U. S. Federal Science database resource (https://www.science.gov/) provides a very helpful access to various scientific reports and publications, including those in animal research and development.

6 References

Abraham, M., McDougald, L. R. and Beckstead, R. B. 2014. Blackhead disease: reduced sensitivity of *Histomonas meleagridis* to nitarsone *in vitro* and *in vivo*. *Avian. Dis.*, 58, 60-3.

Adams, C., Vahl, H. A. and Veldman, A. 1996. Interaction between nutrition and *Eimeria acervulina* infection in broiler chickens: development of an experimental infection model. *Br. J. Nutr.*, 75, 867-73.

Adedokun, S. A. and Olojede, O. C. 2018. Optimizing gastrointestinal integrity in poultry: the role of nutrients and feed additives. *Front. Vet. Sci.*, 5, 348.

Adil, S., Banday, T., Bhat, G. A., Mir, M. S. and Rehman, M. 2010. Effect of dietary supplementation of organic acids on performance, intestinal histomorphology, and serum biochemistry of broiler chicken. *Vet. Med. Int.*, 2010, 479485.

Agunos, A., Carson, C. and Leger, D. 2013. Antimicrobial therapy of selected diseases in turkeys, laying hens, and minor poultry species in Canada. *Can. Vet. J.*, 54, 1041-52.

Akbari, M. R., Haghighi, H. R., Chambers, J. R., Brisbin, J., Read, L. R. and Sharif, S. 2008. Expression of antimicrobial peptides in cecal tonsils of chickens treated with probiotics and infected with *Salmonella enterica* serovar typhimurium. *Clin. Vaccine Immunol.*, 15, 1689-93.

Al-Sheikhly, F. and Truscott, R. B. 1977. The pathology of necrotic enteritis of chickens following infusion of crude toxins of *Clostridium perfringens* into the duodenum. *Avian Dis.*, 21, 241-55.

Alagawany, M., Abd El-Hack, M. E., Farag, M. R., Sachan, S., Karthik, K. and Dhama, K. 2018. The use of probiotics as eco-friendly alternatives for antibiotics in poultry nutrition. *Environ. Sci. Pollut. Res. Int.*, 25, 10611-18.

Alali, W. Q., Hofacre, C. L., Mathis, G. F. and Faltys, G. 2013. Effect of essential oil compound on shedding and colonization of *Salmonella enterica* serovar Heidelberg in broilers. *Poult. Sci.*, 92, 836-41.

Aliakbarpour, H. R., Chamani, M., Rahimi, G., Sadeghi, A. A. and Qujeq, D. 2012. The *Bacillus subtilis* and lactic acid bacteria probiotics influences intestinal mucin gene expression, histomorphology and growth performance in broilers. *Asian-Australas. J. Anim. Sci.*, 25, 1285-93.

Annett, C. B., Viste, J. R., Chirino-Trejo, M., Classen, H. L., Middleton, D. M. and Simko, E. 2002. Necrotic enteritis: effect of barley, wheat and corn diets on proliferation of *Clostridium perfringens* type A. *Avian Pathol.*, 31, 598-601.

Bach Knudsen, K. E., Norskov, N. P., Bolvig, A. K., Hedemann, M. S. and Laerke, H. N. 2017. Dietary fibers and associated phytochemicals in cereals. *Mol. Nutr. Food Res.*, 61.

Barbour, E. K., Bragg, R. R., Karrouf, G., Iyer, A., Azhar, E., Harakeh, S. and Kumosani, T. 2015. Control of eight predominant *Eimeria* spp. involved in economic coccidiosis of broiler chicken by a chemically characterized essential oil. *J. Appl. Microbiol.*, 118, 583-91.

Barnes, D. M., Kirby, Y. K. and Oliver, K. G. 2001. Effects of biogenic amines on growth and the incidence of proventricular lesions in broiler chickens. *Poult. Sci.*, 80, 906-11.

Barrow, P. A., Jones, M. A., Smith, A. L. and Wigley, P. 2012. The long view: *Salmonella*–the last forty years. *Avian Pathol.*, 41, 413-20.

Beach, N. M., Duncan, R. B., Larsen, C. T., Meng, X. J., Sriranganathan, N. and Pierson, F. W. 2009. Persistent infection of turkeys with an avirulent strain of turkey hemorrhagic enteritis virus. *Avian Dis.*, 53, 370-5.

Becerril, R., Nerin, C. and Gomez-Lus, R. 2012. Evaluation of bacterial resistance to essential oils and antibiotics after exposure to oregano and cinnamon essential oils. *Foodborne Pathog. Dis.*, 9, 699-705.

Benavente, J. and Martinez-Costas, J. 2007. Avian reovirus: structure and biology. *Virus Res.*, 123, 105-19.

Berchieri Jr., A., Murphy, C. K., Marston, K. and Barrow, P. A. 2001. Observations on the persistence and vertical transmission of *Salmonella enterica* serovars Pullorum and Gallinarum in chickens: effect of bacterial and host genetic background. *Avian Pathol.*, 30, 221-31.

Beski, S. S. M., Swick, R. A. and Iji, P. A. 2015. Specialized protein products in broiler chicken nutrition: a review. *Anim. Nutr.*, 1, 47-53.

Bourassa, D. V., Wilson, K. M., Ritz, C. R., Kiepper, B. K. and Buhr, R. J. 2018. Evaluation of the addition of organic acids in the feed and/or water for broilers and the subsequent recovery of *Salmonella* Typhimurium from litter and ceca. *Poult. Sci.*, 97, 64-73.

Bradley, R. E. and Radhakrishnan, C. V. 1973. Coccidiosis in chickens: obligate relationship between *Eimeria tenella* and certain species of cecal microflora in the pathogenesis of the disease. *Avian Dis.*, 17, 461-76.

Brisbin, J. T., Gong, J. and Sharif, S. 2008. Interactions between commensal bacteria and the gut-associated immune system of the chicken. *Anim. Health Res. Rev.*, 9, 101-10.

Brisbin, J. T., Gong, J., Parvizi, P. and Sharif, S. 2010. Effects of lactobacilli on cytokine expression by chicken spleen and cecal tonsil cells. *Clin. Vaccine Immunol.*, 17, 1337-43.

Brisbin, J. T., Gong, J., Orouji, S., Esufali, J., Mallick, A. I., Parvizi, P., Shewen, P. E. and Sharif, S. 2011. Oral treatment of chickens with lactobacilli influences elicitation of immune responses. *Clin. Vaccine Immunol.*, 18, 1447-55.

Brisbin, J. T., Parvizi, P. and Sharif, S. 2012. Differential cytokine expression in T-cell subsets of chicken caecal tonsils co-cultured with three species of *Lactobacillus*. *Benef. Microbes*, 3, 205-10.

Brisbin, J. T., Davidge, L., Roshdieh, A. and Sharif, S. 2015. Characterization of the effects of three Lactobacillus species on the function of chicken macrophages. *Res. Vet. Sci.*, 100, 39-44.

Burt, S. 2004. Essential oils: their antibacterial properties and potential applications in foods—a review. *Int. J. Food Microbiol.*, 94, 223-53.

Calnek, B. W. 2015. Avian diseases: the creation and evolution of P. Philip Levine's enduring gift. *Avian Dis.*, 59, 1-6.

Caly, D. L., Chevalier, M., Flahaut, C., Cudennec, B., Al Atya, A. K., Chataigne, G., D'inca, R., Auclair, E. and Drider, D. 2017. The safe enterocin Dd14 is a leaderless two-peptide bacteriocin with anti-*Clostridium perfringens* activity. *Int. J. Antimicrob. Agents*, 49, 282-9.

Carre, B., Gomez, J. and Chagneau, A. M. 1995. Contribution of oligosaccharide and polysaccharide digestion, and excreta losses of lactic acid and short chain fatty acids, to dietary metabolisable energy values in broiler chickens and adult cockerels. *Br. Poult. Sci.*, 36, 611-29.

Chapman, H. D., Barta, J. R., Blake, D., Gruber, A., Jenkins, M., Smith, N. C., Suo, X. and Tomley, F. M. 2013. A selective review of advances in coccidiosis research. *Adv. Parasitol.*, 83, 93-171.

Chen, C. Y., Tsen, H. Y., Lin, C. L., Yu, B. and Chen, C. S. 2012. Oral administration of a combination of select lactic acid bacteria strains to reduce the *Salmonella* invasion and inflammation of broiler chicks. *Poult. Sci.*, 91, 2139-47.

Choct, M., Hughes, R. J., Trimble, R. P., Angkanaporn, K. and Annison, G. 1995. Non-starch polysaccharide-degrading enzymes increase the performance of broiler chickens fed wheat of low apparent metabolizable energy. *J. Nutr.*, 125, 485-92.

Choct, M., Hughes, R. J., Wang, J., Bedford, M. R., Morgan, A. J. and Annison, G. 1996. Increased small intestinal fermentation is partly responsible for the anti-nutritive activity of non-starch polysaccharides in chickens. *Br. Poult. Sci.*, 37, 609-21.

Clavijo, V. and Florez, M. J. V. 2018. The gastrointestinal microbiome and its association with the control of pathogens in broiler chicken production: a review. *Poult. Sci.*, 97, 1006-21.

Cortez, V., Meliopoulos, V. A., Karlsson, E. A., Hargest, V., Johnson, C. and Schultz-Cherry, S. 2017. Astrovirus biology and pathogenesis. *Annu. Rev. Virol.*, 4, 327-48.

Cosentino, S., Tuberoso, C. I., Pisano, B., Satta, M., Mascia, V., Arzedi, E. and Palmas, F. 1999. In-vitro antimicrobial activity and chemical composition of Sardinian *Thymus* essential oils. *Lett. Appl. Microbiol.*, 29, 130-5.

Craven, S. E., Cox, N. A., Bailey, J. S. and Cosby, D. E. 2003. Incidence and tracking of *Clostridium perfringens* through an integrated broiler chicken operation. *Avian. Dis.*, 47, 707-11.

Dabard, J., Bridonneau, C., Phillipe, C., Anglade, P., Molle, D., Nardi, M., Ladire, M., Girardin, H., Marcille, F., Gomez, A. and Fons, M. 2001. Ruminococcin A, a new lantibiotic produced by a *Ruminococcus gnavus* strain isolated from human feces. *Appl. Environ. Microbiol.*, 67, 4111-8.

Dahiya, J. P., Hoehler, D., Wilkie, D. C., Van Kessel, A. G. and Drew, M. D. 2005. Dietary glycine concentration affects intestinal *Clostridium perfringens* and lactobacilli populations in broiler chickens. *Poult. Sci.*, 84, 1875-85.

Dahiya, J. P., Hoehler, D., Van Kessel, A. G. and Drew, M. D. 2007. Effect of different dietary methionine sources on intestinal microbial populations in broiler chickens. *Poult. Sci.*, 86, 2358-66.

Dalloul, R. A., Lillehoj, H. S., Shellem, T. A. and Doerr, J. A. 2003. Intestinal immunomodulation by vitamin A deficiency and lactobacillus-based probiotic in *Eimeria acervulina*-infected broiler chickens. *Avian. Dis.*, 47, 1313-20.

Delaquis, P. J., Stanich, K., Girard, B. and Mazza, G. 2002. Antimicrobial activity of individual and mixed fractions of dill, cilantro, coriander and eucalyptus essential oils. *Int. J. Food. Microbiol.*, 74, 101-9.

Dhama, K., Saminathan, M., Karthik, K., Tiwari, R., Shabbir, M. Z., Kumar, N., Malik, Y. S. and Singh, R. K. 2015. Avian rotavirus enteritis – an updated review. *Vet. Q.*, 35, 142-58.

Dhillon, A. S., Roy, P., Lauerman, L., Schaberg, D., Weber, S., Bandli, D. and Wier, F. 2004. High mortality in egg layers as a result of necrotic enteritis. *Avian. Dis.*, 48, 675-80.

Diarra, M. S. and Malouin, F. 2014. Antibiotics in Canadian poultry productions and anticipated alternatives. *Front. Microbiol.*, 5, 282.

Dittoe, D. K., Ricke, S. C. and Kiess, A. S. 2018. Organic acids and potential for modifying the avian gastrointestinal tract and reducing pathogens and disease. *Front. Vet. Sci.*, 5, 216.

Drew, M. D., Syed, N. A., Goldade, B. G., Laarveld, B. and Van Kessel, A. G. 2004. Effects of dietary protein source and level on intestinal populations of *Clostridium perfringens* in broiler chickens. *Poult. Sci.*, 83, 414–20.

Elwinger, K., Schneitz, C., Berndtson, E., Fossum, O., Teglof, B. and Engstom, B. 1992. Factors affecting the incidence of necrotic enteritis, caecal carriage of *Clostridium perfringens* and bird performance in broiler chicks. *Acta. Vet. Scand.*, 33, 369–78.

Engberg, R. M., Hedemann, M. S., Steenfeldt, S. and Jensen, B. B. 2004. Influence of whole wheat and xylanase on broiler performance and microbial composition and activity in the digestive tract. *Poult. Sci.*, 83, 925–38.

Faber, T. A., Dilger, R. N., Hopkins, A. C., Price, N. P. and Fahey Jr., G. C. 2012. Effects of oligosaccharides in a soybean meal-based diet on fermentative and immune responses in broiler chicks challenged with *Eimeria acervulina*. *Poult. Sci.*, 91, 3132–40.

Farhadi, D., Karimi, A., Sadeghi, G., Sheikhahmadi, A., Habibian, M., Raei, A. and Sobhani, K. 2017. Effects of using eucalyptus (*Eucalyptusglobulus* L.) leaf powder and its essential oil on growth performance and immune response of broiler chickens. *Iran. J. Vet. Res.*, 18, 60–2.

Fatoba, A. J. and Adeleke, M. A. 2018. Diagnosis and control of chicken coccidiosis: a recent update. *J. Parasit. Dis.*, 42, 483–93.

Fernandez-Rubio, C., Ordonez, C., Abad-Gonzalez, J., Garcia-Gallego, A., Honrubia, M. P., Mallo, J. J. and Balana-Fouce, R. 2009. Butyric acid-based feed additives help protect broiler chickens from *Salmonella* Enteritidis infection. *Poult. Sci.*, 88, 943–8.

Ferrandis Vila, M., Trudeau, M. P., Hung, Y. T., Zeng, Z., Urriola, P. E., Shurson, G. C. and Saqui-Salces, M. 2018. Dietary fiber sources and non-starch polysaccharide-degrading enzymes modify mucin expression and the immune profile of the swine ileum. *PLoS One*, 13, e0207196.

Fossum, O., Sandstedt, K. and Engstrom, B. E. 1988. Gizzard erosions as a cause of mortality in White Leghorn chickens. *Avian Pathol.*, 17, 519–25.

Gadde, U., Kim, W. H., Oh, S. T. and Lillehoj, H. S. 2017. Alternatives to antibiotics for maximizing growth performance and feed efficiency in poultry: a review. *Anim. Health. Res. Rev.*, 18, 26–45.

Garozzo, A., Timpanaro, R., Bisignano, B., Furneri, P. M., Bisignano, G. and Castro, A. 2009. *In vitro* antiviral activity of *Melaleuca alternifolia* essential oil. *Lett. Appl. Microbiol.*, 49, 806–8.

Gazdzinski, P. and Julian, R. J. 1992. Necrotic enteritis in turkeys. *Avian. Dis.*, 36, 792–8.

Gelli, A., Becquey, E., Ganaba, R., Headey, D., Hidrobo, M., Huybregts, L., Verhoef, H., Kenfack, R., Zongouri, S. and Guedenet, H. 2017. Improving diets and nutrition through an integrated poultry value chain and nutrition intervention (Selever) in Burkina Faso: study protocol for a randomized trial. *Trials*, 18, 412.

Giannenas, I., Papadopoulos, E., Tsalie, E., Triantafillou, E., Henikl, S., Teichmann, K. and Tontis, D. 2012. Assessment of dietary supplementation with probiotics on performance, intestinal morphology and microflora of chickens infected with *Eimeria tenella*. *Vet. Parasitol.*, 188, 31–40.

Giannenas, I., Tsalie, E., Triantafillou, E., Hessenberger, S., Teichmann, K., Mohnl, M. and Tontis, D. 2014. Assessment of probiotics supplementation via feed or water on the growth performance, intestinal morphology and microflora of chickens after experimental infection with *Eimeria acervulina*, *Eimeria maxima* and *Eimeria tenella*. *Avian Pathol.*, 43, 209–16.

Gilbert, M. S., Ijssennagger, N., Kies, A. K. and Van Mil, S. W. C. 2018. Protein fermentation in the gut; implications for intestinal dysfunction in humans, pigs, and poultry. *Am. J. Physiol. Gastrointest. Liver Physiol.*, 315, G159–70.

Gomathi, G., Senthilkumar, S., Natarajan, A., Amutha, R. and Purushothaman, M. R. 2018. Effect of dietary supplementation of cinnamon oil and sodium butyrate on carcass characteristics and meat quality of broiler chicken. *Vet. World*, 11, 959–64.

Gonzalez-Alvarado, J. M., Jimenez-Moreno, E., Valencia, D. G., Lazaro, R. and Mateos, G. G. 2008. Effects of fiber source and heat processing of the cereal on the development and pH of the gastrointestinal tract of broilers fed diets based on corn or rice. *Poult. Sci.*, 87, 1779–95.

Goossens, E., Valgaeren, B. R., Pardon, B., Haesebrouck, F., Ducatelle, R., Deprez, P. R. and Van Immerseel, F. 2017. Rethinking the role of alpha toxin in *Clostridium perfringens*-associated enteric diseases: a review on bovine necro-haemorrhagic enteritis. *Vet. Res.*, 48, 9.

Greco, M. V., Franchi, M. L., Rico Golba, S. L., Pardo, A. G. and Pose, G. N. 2014. Mycotoxins and mycotoxigenic fungi in poultry feed for food-producing animals. *Sci World J.*, 2014, 968215.

Guerre, P. 2016. Worldwide mycotoxins exposure in pig and poultry feed formulations. *Toxins (Basel)*, 8.

Gulluce, M., Sokmen, M., Daferera, D., Agar, G., Ozkan, H., Kartal, N., Polissiou, M., Sokmen, A. and Sahin, F. 2003. *In vitro* antibacterial, antifungal, and antioxidant activities of the essential oil and methanol extracts of herbal parts and callus cultures of *Satureja hortensis* L. *J. Agric. Food. Chem.*, 51, 3958–65.

Guo, S., Liu, D., Zhang, B., Li, Z., Li, Y., Ding, B. and Guo, Y. 2017. Two *Lactobacillus* species inhibit the growth and alpha-toxin production of *Clostridium perfringens* and induced proinflammatory factors in chicken intestinal epithelial cells *in vitro*. *Front. Microbiol.*, 8, 2081.

Guy, J. S. 1998. Virus infections of the gastrointestinal tract of poultry. *Poult. Sci.*, 77, 1166–75.

Haghighi, H. R., Gong, J., Gyles, C. L., Hayes, M. A., Sanei, B., Parvizi, P., Gisavi, H., Chambers, J. R. and Sharif, S. 2005. Modulation of antibody-mediated immune response by probiotics in chickens. *Clin. Diagn. Lab. Immunol.*, 12, 1387–92.

Haghighi, H. R., Gong, J., Gyles, C. L., Hayes, M. A., Zhou, H., Sanei, B., Chambers, J. R. and Sharif, S. 2006. Probiotics stimulate production of natural antibodies in chickens. *Clin. Vaccine. Immunol.*, 13, 975–80.

Haghighi, H. R., Abdul-Careem, M. F., Dara, R. A., Chambers, J. R. and Sharif, S. 2008. Cytokine gene expression in chicken cecal tonsils following treatment with probiotics and *Salmonella* infection. *Vet. Microbiol.*, 126, 225–33.

Haiko, J. and Westerlund-Wikstrom, B. 2013. The role of the bacterial flagellum in adhesion and virulence. *Biology (Basel)*, 2, 1242–67.

Hernandez, F., Garcia, V., Madrid, J., Orengo, J., Catala, P. and Megias, M. D. 2006. Effect of formic acid on performance, digestibility, intestinal histomorphology and plasma metabolite levels of broiler chickens. *Br. Poult. Sci.*, 47, 50–6.

Hofacre, C. L., Smith, J. A. and Mathis, G. F. 2018. An optimist's view on limiting necrotic enteritis and maintaining broiler gut health and performance in today's marketing, food safety, and regulatory climate. *Poult. Sci.*, 97, 1929–33.

Hughes, R. A., Ali, R. A., Mendoza, M. A., Hassan, H. M. and Koci, M. D. 2017. Impact of dietary galacto-oligosaccharide (Gos) on chicken's gut microbiota, mucosal gene expression, and *Salmonella* colonization. *Front. Vet. Sci.*, 4, 192.

Hume, M. E., Clemente-Hernandez, S. and Oviedo-Rondon, E. O. 2006. Effects of feed additives and mixed *Eimeria* species infection on intestinal microbial ecology of broilers. *Poult. Sci.*, 85, 2106–11.

Jansson, D. S., Nyman, A., Vagsholm, I., Christensson, D., Goransson, M., Fossum, O. and Hoglund, J. 2010. Ascarid infections in laying hens kept in different housing systems. *Avian Pathol.*, 39, 525–32.

Jitviriyanon, S., Phanthong, P., Lomarat, P., Bunyapraphatsara, N., Porntrakulpipat, S. and Paraksa, N. 2016. *In vitro* study of anti-coccidial activity of essential oils from indigenous plants against *Eimeria tenella*. *Vet. Parasitol.*, 228, 96–102.

Jozefiak, D., Sip, A., Rutkowski, A., Rawski, M., Kaczmarek, S., Wolun-Cholewa, M., Engberg, R. M. and Hojberg, O. 2012. Lyophilized *Carnobacterium divergens* As7 bacteriocin preparation improves performance of broiler chickens challenged with *Clostridium perfringens*. *Poult. Sci.*, 91, 1899–907.

Kaingu, F., Liu, D., Wang, L., Tao, J., Waihenya, R. and Kutima, H. 2017. Anticoccidial effects of *Aloe secundiflora* leaf extract against *Eimeria tenella* in broiler chicken. *Trop. Anim. Health. Prod.*, 49, 823–8.

Kalemba, D. and Kunicka, A. 2003. Antibacterial and antifungal properties of essential oils. *Curr. Med. Chem.*, 10, 813–29.

Keerqin, C., Morgan, N. K., Wu, S. B., Swick, R. A. and Choct, M. 2017. Dietary inclusion of arabinoxylo-oligosaccharides in response to broilers challenged with subclinical necrotic enteritis. *Br. Poult. Sci.*, 58, 418–24.

Kiarie, E. G. and Mills, A. 2019. Role of feed processing on gut health and function in pigs and poultry: conundrum of optimal particle size and hydrothermal regimens. *Front. Vet. Sci.*, 6, 19.

Kim, G. B., Seo, Y. M., Kim, C. H. and Paik, I. K. 2011. Effect of dietary prebiotic supplementation on the performance, intestinal microflora, and immune response of broilers. *Poult. Sci.*, 90, 75–82.

Kim, S. J., Lee, K. W., Kang, C. W. and An, B. K. 2016. Growth performance, relative meat and organ weights, cecal microflora, and blood characteristics in broiler chickens fed diets containing different nutrient density with or without essential oils. *Asian-Australas. J. Anim. Sci.*, 29, 549–54.

Korver, D. R., Zuidhof, M. J. and Lawes, K. R. 2004. Performance characteristics and economic comparison of broiler chickens fed wheat- and triticale-based diets. *Poult. Sci.*, 83, 716–25.

Koynarski, V., Stoev, S., Grozeva, N., Mirtcheva, T., Daskalov, H., Mitev, J. and Mantle, P. 2007. Experimental coccidiosis provoked by *Eimeria acervulina* in chicks simultaneously fed on ochratoxin A contaminated diet. *Res. Vet. Sci.*, 82, 225–31.

Koyuncu, S., Andersson, M. G., Lofstrom, C., Skandamis, P. N., Gounadaki, A., Zentek, J. and Haggblom, P. 2013. Organic acids for control of *Salmonella* in different feed materials. *Bmc. Vet. Res.*, 9, 81.

La Ragione, R. M., Narbad, A., Gasson, M. J. and Woodward, M. J. 2004. *In vivo* characterization of *Lactobacillus johnsonii* Fi9785 for use as a defined competitive exclusion agent against bacterial pathogens in poultry. *Lett. Appl. Microbiol.*, 38, 197–205.

Langhout, D. J., Schutte, J. B., De Jong, J., Sloetjes, H., Verstegen, M. W. and Tamminga, S. 2000. Effect of viscosity on digestion of nutrients in conventional and germ-free chicks. *Br. J. Nutr.*, 83, 533–40.

Leeson, S., Zubair, A. K., Squires, E. J. and Forsberg, C. 1997. Influence of dietary levels of fat, fiber, and copper sulfate and fat rancidity on cecal activity in the growing turkey. *Poult. Sci.*, 76, 59–66.

Lensing, M., Van Der Klis, J. D., Yoon, I. and Moore, D. T. 2012. Efficacy of *Saccharomyces cerevisiae* fermentation product on intestinal health and productivity of coccidian-challenged laying hens. *Poult. Sci.*, 91, 1590–7.

Lepp, D., Roxas, B., Parreira, V. R., Marri, P. R., Rosey, E. L., Gong, J., Songer, J. G., Vedantam, G. and Prescott, J. F. 2010. Identification of novel pathogenicity loci in *Clostridium perfringens* strains that cause avian necrotic enteritis. *PLoS One*, 5, e10795.

Li, J., Wang, W., Xu, S. X., Magarvey, N. A. and Mccormick, J. K. 2011. *Lactobacillus reuteri*-produced cyclic dipeptides quench agr-mediated expression of toxic shock syndrome toxin-1 in staphylococci. *Proc. Natl. Acad. Sci. U. S. A.*, 108, 3360–5.

Lin, Y., Xu, S., Zeng, D., Ni, X., Zhou, M., Zeng, Y., Wang, H., Zhou, Y., Zhu, H., Pan, K. and Li, G. 2017. Disruption in the cecal microbiota of chickens challenged with *Clostridium perfringens* and other factors was alleviated by *Bacillus licheniformis* supplementation. *PLoS One*, 12, e0182426.

Lovland, A. and Kaldhusdal, M. 1999. Liver lesions seen at slaughter as an indicator of necrotic enteritis in broiler flocks. *Fems. Immunol. Med. Microbiol.*, 24, 345–51.

M'Sadeq, S. A., Wu, S. B., Swick, R. A. and Choct, M. 2015a. Towards the control of necrotic enteritis in broiler chickens with in-feed antibiotics phasing-out worldwide. *Anim. Nutr.*, 1, 1–11.

M'Sadeq, S. A., Wu, S. B., Choct, M., Forder, R. and Swick, R. A. 2015b. Use of yeast cell wall extract as a tool to reduce the impact of necrotic enteritis in broilers. *Poult. Sci.*, 94, 898–905.

Maisonnier, S., Gomez, J., Bree, A., Berri, C., Baeza, E. and Carre, B. 2003. Effects of microflora status, dietary bile salts and guar gum on lipid digestibility, intestinal bile salts, and histomorphology in broiler chickens. *Poult. Sci.*, 82, 805–14.

Manafi, M., Hedayati, M., Pirany, N. and Omede, A. A. 2019. Comparison of performance and feed digestibility of the non-antibiotic feed supplement (Novacid) and an antibiotic growth promoter in broiler chickens. *Poult. Sci.*, 98, 904–11.

Manes-Lazaro, R., Van Diemen, P. M., Pin, C., Mayer, M. J., Stevens, M. P. and Narbad, A. 2017. Administration of *Lactobacillus johnsonii* Fi9785 to chickens affects colonisation by *Campylobacter jejuni* and the intestinal microbiota. *Br. Poult. Sci.*, 58, 373–81.

Marotta, F., Garofolo, G., Di Donato, G., Aprea, G., Platone, I., Cianciavicchia, S., Alessiani, A. and Di Giannatale, E. 2015. Population diversity of *Campylobacter jejuni* in poultry and its dynamic of contamination in chicken meat. *Biomed. Res. Int.*, 2015, 859845.

McDougald, L. R. and Fuller, L. 2005. Blackhead disease in turkeys: direct transmission of *Histomonas meleagridis* from bird to bird in a laboratory model. *Avian. Dis.*, 49, 328–31.

McDougald, L. R., Abraham, M. and Beckstead, R. B. 2012. An outbreak of blackhead disease (*Histomonas meleagridis*) in farm-reared bobwhite quail (*Colinus virginianus*). *Avian. Dis.*, 56, 754–6.

Mehde, A. A., Mehdi, W. A., Ozacar, M. and Ozacar, Z. Z. 2018. Evaluation of different saccharides and chitin as eco-friendly additive to improve the magnetic cross-linked enzyme aggregates (CLEAs) activities. *Int. J. Biol. Macromol.*, 118, 2040–50.

Melo, A. D., Amaral, A. F., Schaefer, G., Luciano, F. B., De Andrade, C., Costa, L. B. and Rostagno, M. H. 2015. Antimicrobial effect against different bacterial strains and bacterial adaptation to essential oils used as feed additives. *Can. J. Vet. Res.*, 79, 285–9.

Mettifogo, E., Nunez, L. F., Chacon, J. L., Santander Parra, S. H., Astolfi-Ferreira, C. S., Jerez, J. A., Jones, R. C. and Piantino Ferreira, A. J. 2014. Emergence of enteric viruses in production chickens is a concern for avian health. *Sci. World J.*, 2014, 450423.

Mikkelsen, L. L., Vidanarachchi, J. K., Olnood, C. G., Bao, Y. M., Selle, P. H. and Choct, M. 2009. Effect of potassium diformate on growth performance and gut microbiota in broiler chickens challenged with necrotic enteritis. *Br. Poult. Sci.*, 50, 66–75.

Mith, H., Dure, R., Delcenserie, V., Zhiri, A., Daube, G. and Clinquart, A. 2014. Antimicrobial activities of commercial essential oils and their components against food-borne pathogens and food spoilage bacteria. *Food. Sci. Nutr.*, 2, 403–16.

Mitsch, P., Zitterl-Eglseer, K., Kohler, B., Gabler, C., Losa, R. and Zimpernik, I. 2004. The effect of two different blends of essential oil components on the proliferation of *Clostridium perfringens* in the intestines of broiler chickens. *Poult. Sci.*, 83, 669–75.

Moore, R. J. 2016. Necrotic enteritis predisposing factors in broiler chickens. *Avian Pathol.*, 45, 275–81.

Morrissey, K. L., Widowski, T., Leeson, S., Sandilands, V., Arnone, A. and Torrey, S. 2014. The effect of dietary alterations during rearing on growth, productivity, and behavior in broiler breeder females. *Poult. Sci.*, 93, 285–95.

Mot, D., Timbermont, L., Haesebrouck, F., Ducatelle, R. and Van Immerseel, F. 2014. Progress and problems in vaccination against necrotic enteritis in broiler chickens. *Avian Pathol.*, 43, 290–300.

Muyyarikkandy, M. S. and Amalaradjou, M. A. 2017. *Lactobacillus bulgaricus*, *Lactobacillus rhamnosus* and *Lactobacillus paracasei* attenuate *Salmonella* Enteritidis, *Salmonella* Heidelberg and *Salmonella* Typhimurium colonization and virulence gene expression *in vitro*. *Int. J. Mol. Sci.*, 18.

Najarian, A., Sharif, S. and Griffiths, M. W. 2019. Evaluation of protective effect of *Lactobacillus acidophilus* La-5 on toxicity and colonization of *Clostridium difficile* in human epithelial cells *in vitro*. *Anaerobe*, 55, 142–51.

Nakphaichit, M., Thanomwongwattana, S., Phraephaisarn, C., Sakamoto, N., Keawsompong, S., Nakayama, J. and Nitisinprasert, S. 2011. The effect of including *Lactobacillus reuteri* Kub-Ac5 during post-hatch feeding on the growth and ileum microbiota of broiler chickens. *Poult. Sci.*, 90, 2753–65.

Namkung, H., Li, M., Gong, J., Yu, H., Cottrill, M. and De Lange, C. F. M. 2004. Impact of feeding blends of organic acids and herbal extracts on growth performance, gut microbiota and digestive function in newly weaned pigs. *Can J Anim Sci*, 84, 697–704.

Oviedo-Rondon, E. O., Hume, M. E., Hernandez, C. and Clemente-Hernandez, S. 2006. Intestinal microbial ecology of broilers vaccinated and challenged with mixed *Eimeria* species, and supplemented with essential oil blends. *Poult. Sci.*, 85, 854–60.

Paiva, D. M., Walk, C. L. and Mcelroy, A. P. 2013. Influence of dietary calcium level, calcium source, and phytase on bird performance and mineral digestibility during a natural necrotic enteritis episode. *Poult. Sci.*, 92, 3125–33.

Palliyeguru, M. W., Rose, S. P. and Mackenzie, A. M. 2010. Effect of dietary protein concentrates on the incidence of subclinical necrotic enteritis and growth performance of broiler chickens. *Poult. Sci.*, 89, 34–43.

Palya, V., Nagy, M., Glavits, R., Ivanics, E., Szalay, D., Dan, A., Suveges, T., Markos, B. and Harrach, B. 2007. Investigation of field outbreaks of turkey haemorrhagic enteritis in Hungary. *Acta. Vet. Hung.*, 55, 135–49.

Pan, D. and Yu, Z. 2014. Intestinal microbiome of poultry and its interaction with host and diet. *Gut Microbes*, 5, 108–19.

Pascual, M., Hugas, M., Badiola, J. I., Monfort, J. M. and Garriga, M. 1999. *Lactobacillus salivarius* CTC2197 prevents *Salmonella enteritidis* colonization in chickens. *Appl. Environ. Microbiol.*, 65, 4981–6.

Persia, M. E., Young, E. L., Utterback, P. L. and Parsons, C. M. 2006. Effects of dietary ingredients and *Eimeria acervulina* infection on chick performance, apparent metabolizable energy, and amino acid digestibility. *Poult. Sci.*, 85, 48–55.

Pessoa, L. M., Morais, S. M., Bevilaqua, C. M. and Luciano, J. H. 2002. Anthelmintic activity of essential oil of *Ocimum gratissimum* Linn. and eugenol against *Haemonchus contortus*. *Vet. Parasitol.*, 109, 59–63.

Pourabedin, M. and Zhao, X. 2015. Prebiotics and gut microbiota in chickens. *Fems. Microbiol. Lett.*, 362, fnv122.

Powers, W. and Capelari, M. 2017. Production, management and the environment symposium: measurement and mitigation of reactive nitrogen species from swine and poultry production. *J. Anim. Sci.*, 95, 2236–40.

Prescott, J. F., Parreira, V. R., Mehdizadeh Gohari, I., Lepp, D. and Gong, J. 2016. The pathogenesis of necrotic enteritis in chickens: what we know and what we need to know: a review. *Avian Pathol.*, 45, 288–94.

Ribet, D. and Cossart, P. 2015. How bacterial pathogens colonize their hosts and invade deeper tissues. *Microbes Infect.*, 17, 173–83.

Ricke, S. C. 2018. Impact of prebiotics on poultry production and food safety. *Yale. J. Biol. Med.*, 91, 151–9.

Ritzi, M. M., Abdelrahman, W., Mohnl, M. and Dalloul, R. A. 2014. Effects of probiotics and application methods on performance and response of broiler chickens to an *Eimeria* challenge. *Poult. Sci.*, 93, 2772–8.

Ritzi, M. M., Abdelrahman, W., Van-Heerden, K., Mohnl, M., Barrett, N. W. and Dalloul, R. A. 2016. Combination of probiotics and coccidiosis vaccine enhances protection against an *Eimeria* challenge. *Vet. Res.*, 47, 111.

Rodjan, P., Soisuwan, K., Thongprajukaew, K., Theapparat, Y., Khongthong, S., Jeenkeawpieam, J. and Salaeharae, T. 2018. Effect of organic acids or probiotics alone or in combination on growth performance, nutrient digestibility, enzyme activities, intestinal morphology and gut microflora in broiler chickens. *J. Anim. Physiol. Anim. Nutr (Berl)*, 102, e931–40.

Sakkas, H. and Papadopoulou, C. 2017. Antimicrobial activity of basil, oregano, and thyme essential oils. *J. Microbiol. Biotechnol.*, 27, 429–38.

Salter, A. M. 2017. Improving the sustainability of global meat and milk production. *Proc. Nutr. Soc.*, 76, 22–7.

Shahid, M. Z., Saima, H., Yasmin, A., Nadeem, M. T., Imran, M. and Afzaal, M. 2018. Antioxidant capacity of cinnamon extract for palm oil stability. *Lipids. Health. Dis.*, 17, 116.

Sharma, N., Gupta, A. and Gautam, N. 2014. Characterization of Bacteriocin like inhibitory substance produced by a new Strain *Brevibacillus borstelensis* Ag1 Isolated from 'Marcha'. *Braz. J. Microbiol.*, 45, 1007-15.

Shin, M. S., Han, S. K., Ji, A. R., Kim, K. S. and Lee, W. K. 2008. Isolation and characterization of bacteriocin-producing bacteria from the gastrointestinal tract of broiler chickens for probiotic use. *J. Appl. Microbiol.*, 105, 2203-12.

Shojadoost, B., Vince, A. R. and Prescott, J. F. 2012. The successful experimental induction of necrotic enteritis in chickens by *Clostridium perfringens*: a critical review. *Vet. Res.*, 43, 74.

Sklan, D., Shelly, M., Makovsky, B., Geyra, A., Klipper, E. and Friedman, A. 2003. The effect of chronic feeding of diacetoxyscirpenol and T-2 toxin on performance, health, small intestinal physiology and antibody production in turkey poults. *Br. Poult. Sci.*, 44, 46-52.

Slawinska, A., Plowiec, A., Siwek, M., Jaroszewski, M. and Bednarczyk, M. 2016. Long-term transcriptomic effects of prebiotics and synbiotics delivered *in ovo* in broiler chickens. *PLoS ONE*, 11, e0168899.

Smirnov, A., Perez, R., Amit-Romach, E., Sklan, D. and Uni, Z. 2005. Mucin dynamics and microbial populations in chicken small intestine are changed by dietary probiotic and antibiotic growth promoter supplementation. *J. Nutr.*, 135, 187-92.

Smith-Palmer, A., Stewart, J. and Fyfe, L. 1998. Antimicrobial properties of plant essential oils and essences against five important food-borne pathogens. *Lett. Appl. Microbiol.*, 26, 118-22.

Strompfova, V., Laukova, A., Marcinakova, M. and Vasilkova, Z. 2010. Testing of probiotic and bacteriocin-producing lactic acid bacteria towards *Eimeria* sp. *Pol. J. Vet. Sci.*, 13, 389-91.

Suresh, G., Das, R. K., Kaur Brar, S., Rouissi, T., Avalos Ramirez, A., Chorfi, Y. and Godbout, S. 2018. Alternatives to antibiotics in poultry feed: molecular perspectives. *Crit. Rev. Microbiol.*, 44, 318-35.

Szkotnicki, J. 2013. Antimicrobial therapy of bacterial diseases in broiler chickens – a comment. *Can. Vet. J.*, 54, 201.

Taha-Abdelaziz, K., Hodgins, D. C., Lammers, A., Alkie, T. N. and Sharif, S. 2018. Effects of early feeding and dietary interventions on development of lymphoid organs and immune competence in neonatal chickens: a review. *Vet. Immunol. Immunopathol.*, 201, 1-11.

Teng, P. Y. and Kim, W. K. 2018. Review: roles of prebiotics in intestinal ecosystem of broilers. *Front. Vet. Sci.*, 5, 245.

Teo, A. Y. and Tan, H. M. 2005. Inhibition of Clostridium perfringens by a novel strain of *Bacillus subtilis* isolated from the gastrointestinal tracts of healthy chickens. *Appl. Environ. Microbiol.*, 71, 4185-90.

Tuomola, E., Crittenden, R., Playne, M., Isolauri, E. and Salminen, S. 2001. Quality assurance criteria for probiotic bacteria. *Am. J. Clin. Nutr.*, 73, 393S-8S.

Udompijitkul, P., Paredes-Sabja, D. and Sarker, M. R. 2012. Inhibitory effects of nisin against *Clostridium perfringens* food poisoning and nonfood-borne isolates. *J. Food. Sci.*, 77, M51-6.

Van Coillie, E., Goris, J., Cleenwerck, I., Grijspeerdt, K., Botteldoorn, N., Van Immerseel, F., De Buck, J., Vancanneyt, M., Swings, J., Herman, L. and Heyndrickx, M. 2007. Identification of lactobacilli isolated from the cloaca and vagina of laying hens and characterization for potential use as probiotics to control *Salmonella* Enteritidis. *J. Appl. Microbiol.*, 102, 1095-106.

Van Immerseel, F., Russell, J. B., Flythe, M. D., Gantois, I., Timbermont, L., Pasmans, F., Haesebrouck, F. and Ducatelle, R. 2006. The use of organic acids to combat *Salmonella* in poultry: a mechanistic explanation of the efficacy. *Avian Pathol.*, 35, 182-8.

Van Immerseel, F., Lyhs, U., Pedersen, K. and Prescott, J. F. 2016. Recent breakthroughs have unveiled the many knowledge gaps in *Clostridium perfringens*-associated necrotic enteritis in chickens: the first International Conference on Necrotic Enteritis in Poultry. *Avian Pathol.*, 45, 269-70.

Varel, V. H., Wells, J. E. and Miller, D. N. 2007. Combination of a urease inhibitor and a plant essential oil to control coliform bacteria, odour production and ammonia loss from cattle waste. *J. Appl. Microbiol.*, 102, 472-7.

Wang, S., Peng, Q., Jia, H. M., Zeng, X. F., Zhu, J. L., Hou, C. L., Liu, X. T., Yang, F. J. and Qiao, S. Y. 2017. Prevention of *Escherichia coli* infection in broiler chickens with *Lactobacillus plantarum* B1. *Poult. Sci.*, 96, 2576-86.

Wati, T., Ghosh, T. K., Syed, B. and Haldar, S. 2015. Comparative efficacy of a phytogenic feed additive and an antibiotic growth promoter on production performance, caecal microbial population and humoral immune response of broiler chickens inoculated with enteric pathogens. *Anim. Nutr.*, 1, 213-19.

Weber, D. J., Rutala, W. A., Fischer, W. A., Kanamori, H. and Sickbert-Bennett, E. E. 2016. Emerging infectious diseases: focus on infection control issues for novel coronaviruses (Severe Acute Respiratory Syndrome-CoV and Middle East Respiratory Syndrome-CoV), hemorrhagic fever viruses (Lassa and Ebola), and highly pathogenic avian influenza viruses, A(H5N1) and A(H7N9). *Am. J. Infect. Control.*, 44, e91-100.

Wigley, P., Hulme, S. D., Powers, C., Beal, R. K., Berchieri Jr., A., Smith, A. and Barrow, P. 2005. Infection of the reproductive tract and eggs with *Salmonella enterica* serovar pullorum in the chicken is associated with suppression of cellular immunity at sexual maturity. *Infect. Immun.*, 73, 2986-90.

Williams, R. B. 2005. Intercurrent coccidiosis and necrotic enteritis of chickens: rational, integrated disease management by maintenance of gut integrity. *Avian Pathol.*, 34, 159-80.

Zhai, H., Liu, H., Wang, S., Wu, J. and Kluenter, A. M. 2018. Potential of essential oils for poultry and pigs. *Anim. Nutr.*, 4, 179-86.

Zubair, A. K., Forsberg, C. W. and Leeson, S. 1996. Effect of dietary fat, fiber, and monensin on cecal activity in turkeys. *Poult. Sci.*, 75, 891-9.

Chapter 3

Disease management in poultry flocks

Peter Groves, University of Sydney, Australia

1 Introduction

This chapter will focus on disease preventative measures, health monitoring and disease investigation techniques, and the management of sick, meat chicken flocks. Poultry operations vary markedly, not only between countries, but also between integrated operations within the same geographic region. Diseases in intensive poultry flocks may occur due to viral, bacterial, fungal, protozoal, metazoan or arthropod infections or infestations. Non-infective metabolic conditions, nutritional deficiencies and toxicities can also present disease or poor performance manifestations. Genetic change in the broiler, which has become exceedingly rapid over the last three to four decades, have made major changes to the birds' physiology and anatomy. These changes have often proved unbalanced and resulted in the emergence of disease (typically cardiovascular and skeletal problems) or impaired health states. Technological changes in environmental control of houses, increased stocking densities, shorter batch cycles due to improved growth rate, increased feed intakes, faster turnarounds between batches, growing birds to larger final body weights, multiple pickups for slaughter within a house and regulatory withdrawal of many medications have complicated disease expression and severity. Development of better and expanded vaccines (especially of the immunosuppressive diseases), vaccination technologies and eradication of some diseases from breeding stock have improved bird health and decreased the reliance on antibiotic medication of earlier decades. With shorter grow out time to final weight (now as young as 32 days), the influence of factors during fertile egg incubation assume far greater effects on the growth and health of the

http://dx.doi.org/10.19103/AS.2016.0011.19

modern broiler. As a result, the occurrence of multifactorial conditions has become much more a feature of the scenario and these issues now dominate veterinary interventions and investigations in intensive poultry. The coverage here will tend to be general in nature.

2 Disease investigation techniques

Disease or flock problems may present as either a localized situation (i.e. one farm or one region showing the problem) or as a widespread issue affecting many flocks, possibly in various geographic locations. The latter will require a different approach to the former in terms of complexity and importance to the operation.

2.1 Initial investigation – routine approach

Investigation of a single problem in bird health or poor performance can be approached in the classical veterinary investigation manner: establish the existence of a problem; gather a sound and comprehensive history from the farmer and the service personnel serving the operation, and whatever records of flock performance are available; examine the birds' environment and management (farm area, house conditions, temperature and humidity, air quality (particularly ammonia level), litter quality, ventilation control, biosecurity, vermin control, bird behaviour and clinical appearance of the flock, condition of the range area if applicable); examine fresh dead and culled birds, and culled 'average' birds if necessary by necropsy; and decide on what laboratory specimens to collect (blood, swabs, tissues, histopathology specimens, feed, water, environmental samples) as appropriate to the condition, including antibiotic sensitivity testing in bacterial cases. These areas are well described in standard poultry health texts[1,2].

A diagnosis can usually be generated in such cases and treatment can be instigated if possible, and control measures can be described to prevent the problem in future batches of birds. In integrated operations, however, a localized disease problem is the exception, and usually a problem will be spread across or beyond the operation into the local industry. Where a widespread problem develops, a more systematic approach is required to better understand the situation.

2.2 Widespread problem – systematic approach

When a problem appears to be spread across multiple farms within an operation or across operations in a location, a systematic approach to understanding the situation is a far more rewarding procedure. An epidemiological approach will yield more useful outcomes than a non-systematic empirical investigation. The latter is much more susceptible to bias[3].

A systematic investigation will have more ability to determine problems that have a multifactorial causation, which are the predominant forms of problems seen in modern poultry operations. Where multiple factors contribute to a presenting problem, simple solutions are rare due to the intricacies of possible interactions between the factors[4] and the frequent presence of confounding factors[5] (i.e. factors that may not have a direct bearing on the condition but are associated with other factors that do).

Diseases in poultry are seldom just the result of the presence of an infectious organism. Although a pathogen may be a 'necessary' cause (i.e. its presence is necessary for the disease to occur), it may only be part of a 'sufficient' cause (i.e. a combination of determinants which will always produce the disease). For example, the necessary cause of necrotic enteritis may be *Clostridium perfringens* but the disease will not occur just because that organism is present in the chickens' intestinal tracts but it requires the interaction of several other factors such as coccidiosis, some immunosuppression and excessive protein in the feed (which may make up a sufficient cause) for disease expression.

The objectives of a systematic epidemiological investigation will be to stop the problem from occurring and to enable the prevention of its reoccurrence[6]. The approach is applicable to both infectious and non-infectious diseases and is equally valuable when poor performance is the problem rather than overt disease. The overall aim is to identify factors that can be considered as 'key determinants'[5,6] of the problem (i.e. those factors which are important in increasing or decreasing the risk of occurrence of the problem, which can be influenced by management practices).

Hancock and Wiske (1998)[6] suggested a useful starting point with an as yet misunderstood flock problem to prepare a path model by drawing together known information about the condition. This is not a mathematical model but a descriptive attempt to describe how some risk factors may relate to the situation. This can form the basis of areas to be evaluated within the overall investigation. An example of a path model developed to aid understanding of the broiler ascites syndrome (pulmonary hypertension syndrome) is illustrated in Fig. 1.

The first requirement in this scenario is to define a 'case'. In poultry, the smallest unit of study possible is often the 'house' or even 'farm' rather than individual animals. A 'case' should be defined to describe a 'unit of study', which has a significant enough level of the disease or depressed performance that is typical of the problem, differing from 'healthier' flocks which can be considered as 'controls'.

The distribution of cases across time and spatial location initially may lead to a closer appreciation of the problem's occurrence. Putative risk factors identified from the path modelling procedure or drawn from other contributing sources should be examined in

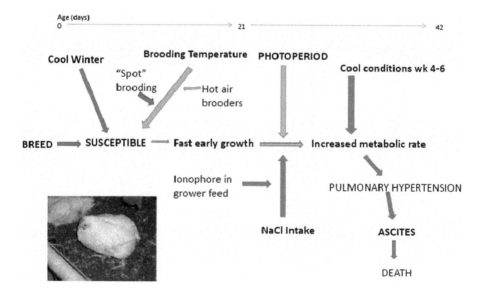

Figure 1 A path model for broiler ascites syndrome. Red arrows represent risk factors and blue arrows represent protective factors (adapted from Groves, 1999[7]).

their association with identified case and control flocks. The use of recorded performance data (see Section 4) can be extremely useful to gain quantitative estimates of the influence of these factors. These may be evaluated using the classical epidemiological contingency table approach (χ^2 analysis) or by t-tests or ANOVA if the independent variable is continuous[3]. Bearing in mind the importance of the health and immunological status of the broilers' breeder source (see Section 4.1), the identity of the birds' parents should always be included as a risk factor to be evaluated. Other factors commonly included in the initial investigation include date of placement, age of birds, timing and pattern of mortality or morbidity or performance loss, sex (if farmed separately), breed and natural groupings such as house types, litter type, age of use of litter, source hatchery, feed source and ration stage, flock size, stocking density at certain ages, vaccination history and history of any medications. Other factors need to be included if data is available and can include any factors thought to be associated (e.g. ventilation levels, water consumption, pickup for slaughter ages, proximity to other poultry farms, proximity to main roads, altitude, feeder and drinker types, water quality and source, interaction of staff with other farms, sharing of equipment with other farms, proximity to surface water bodies and presence of wild birds).

A univariate analysis across each factor's association with cases or controls will yield a shorter list of interesting potential risk factors. The analysis should then move to multivariate techniques, searching for confounding interactions that may modify or clarify the picture. Multivariate techniques such as multiple logistic regressions or multiple linear regressions can be powerful in this process. The significant risk or protective factors should have become apparent or developed into suggestions of other areas to investigate. Interactions are particularly important at this stage.

If findings from the systematic approach lead to new knowledge and perspectives, depending on the importance of the problem and the immediate ability to remedy it, the identified risk factors provide opportunities for directed experimental studies to firstly confirm the association between the putative risk factor(s) and the disease occurrence. This will then lead to more targeted experimental work to understand the disease's pathogenesis and to look more closely at control options. This chronology of approach generally will provide a more focused programme than if experimental studies are launched on a circumstantial and experiential basis. Having identified risk factors will greatly facilitate research and generate more positive outcomes more quickly than a project developed from empirical processes which can be highly susceptible to biases.

Risk factors identified as above can also be swiftly investigated in designed field cohort studies or clinical trials. The results of any experimental work conducted should also be positioned into field evaluations to confirm findings in a commercial context, which often differs markedly from small-scale experimental studies.

This threefold approach: systematic epidemiological investigation yielding putative risk factors, implementation of the risk factors into controlled experiments and confirmation of the findings in field situations is the most powerful and efficient method of understanding and improving control of complex disease and performance problems available.

3 Preventative measures

Best practice disease control measures can decrease the risk of infection and may reduce the severity of an endemic infection. The components of disease prevention and control measures include biosecurity, vaccination, medication and health monitoring.

3.1 Biosecurity

Biosecurity is simply defined as 'security from transmission of infectious diseases, parasites and pests'.[8]

Some infectious diseases are capable of being prevented on farms by good biosecurity while others cannot be completely controlled by this means. For example, biosecurity may be effective against diseases where the pathogen does not survive well outside the host (e.g. mycoplasmata, infectious laryngotracheitis virus), requires relatively close contact for transmission (e.g. infectious coryza) and where there are no alternative control measures (e.g. avian influenza (AI) virus where the jurisdiction does not allow vaccination). Situations where biosecurity alone is likely to be ineffective include those where the pathogen has a long survival time in the environment (e.g. Marek's disease virus and fowl cholera), has an airborne route of infection (e.g. infectious bronchitis virus and Marek's disease virus), has an effective and robust vector (e.g. fowl pox virus), is fairly resistant to disinfectants (e.g. coccidiosis and infectious bursal disease virus) or where the consequences of an outbreak are severe (e.g. infectious laryngotracheitis virus). These latter diseases will require added control mechanisms on top of biosecurity (e.g. prophylactic medication or vaccination) to achieve adequate control.

The basics of biosecurity as practised in the chicken meat industry have several components, as listed below.

3.1.1 Hierarchy of operations

Generally, the effort devoted to biosecurity will vary depending on the class of bird and their relative value and importance to the operation. This may also be a determinant in their ability to spread an infection throughout an operation. For example, the introduction of a *Mycoplasma* into an elite breeder flock could lead to vertical transmission throughout several generations of birds through vertical transmission, whereas the introduction of the same organism into a commercial broiler flock would be self-limiting. Hence many integrated operations would apply strict biosecurity to the elite flocks higher in

Table 1 Example of differing emphases on biosecurity applied to different classes of poultry in some countries

Procedure	Breeders and hatcheries	Broilers	Layers
Isolated	Yes, but often on complexes	No	No
All-in, all-out	Yes	Yes	No
Visitor register	Yes	Yes	Larger operations
Shower in	Yes	No	No
PPE	Yes	Yes	Variable
Foot baths and hand washes	Yes	Yes	Variable
Vehicle disinfection	Yes	Yes/No	Mostly No
Remote feed delivery	Yes	No	Mostly No
Youngest to oldest	Only limited visits	Yes	No
Dedicated staff	Yes	Varies	Varies

the hierarchy (nucleus, great grandparent breeders and grandparents) and somewhat less stringent measures to lower stages (hatcheries, parent breeders) and less still to commercial broiler flocks. Differences in biosecurity application to various levels may relate to the limitations imposed on time allowed between visits of personnel (such as service staff and veterinarians) between farms (e.g. a break in contact between a visit to a great grandparent operation of 48 hours, a shorter break before visiting a grandparent flock of 36 hours, an overnight break before visiting a hatchery but would allow multiple visits to broiler farms within the same day).

An example of different levels of biosecurity procedures commonly applied to different classes of poultry is shown in Table 1.

3.1.2 Isolation of premises

Isolation used in this context refers to distance, in space and time, between chicken operations. 'Isolation' in most biosecurity thinking refers to separation of newly introduced animals for a time to observe whether they are healthy. Where commercial poultry flocks introduce new animals it will usually be a group of many thousands and separation for a time would require major facilities and huge labour requirements in moving them a second time. This is highly impractical in a commercial situation. By 'isolation' in this context refers to the separation of farms by distance – not only from other poultry farms but also from main roads which may carry birds being moved and from personnel who may have contact with other birds, poultry or otherwise. Many chicken diseases are capable of airborne spread or can be readily transmitted by fomites, the most important of which would be people and equipment which may be shared between farms. Marek's disease virus, for example, can spread at least 2–3 km through airborne route[9]. Infectious bronchitis virus is also capable of airborne spread over long distances[10]. Birds being trucked to slaughter or replacement pullets that travel past other poultry operations can be a source of infection to flocks close to main roads (e.g. infectious laryngotracheitis virus and *Mycoplasma gallisepticum*).

3.1.3 Single age

Introduction of an 'all-in, all-out' approach to commercial poultry farming improved health dramatically and provided far better disease exclusion capability. Avoidance of multi-age flocks breaks the cycle of many diseases by interrupting the ability of infectious organisms to spread from older, infected flocks into young naïve birds. This is particularly important for pathogens with long carrier states, where recovered individuals may harbour the organism without clinical signs for long periods. Obvious examples are the mycoplasmata, infectious laryngotracheitis and infectious coryza. Large integrated poultry companies have the capacity to place all houses on a large broiler complex within a week. The established flock can also be depopulated within a short period, allowing for terminal hygiene procedures prior to the arrival of the next batch of broilers.

Breeder flocks also should be managed in an 'all-in, all-out' fashion, with all flocks placed within a few weeks span. Once a breeder flock is established as a single age unit it is desirable to maintain it as a closed flock, with no further introductions during its lifetime. This objective is frequently compromised, however, with the need to introduce fresh males in a middle-aged breeder flock to maintain fertility. This practice is a major compromise to biosecurity and risks introduction of a pathogen with the introduced birds, or exposure of the younger, more naïve males to a resident pathogen in the existing flock.

3.1.4 Quarantine and hygiene procedures

There are many practices which can be considered and the level of adoption on a particular farm will depend on the farm's position in the hierarchy, importance and value to the operation and the management philosophy on biosecurity. Practices may include:

- Provision of an entry log for visitors to the farm.
- Vehicle disinfection at the point of entry to the property. Feed trucks and gas delivery trucks may need to enter the site but also have the propensity to have recently visited other poultry farms during their normal business. Thus they represent a frequent risk for the possible introduction of pathogens. Vehicle disinfection facilities may vary from the simple provision of manually operated disinfectant sprays (Fig. 2a) to more sophisticated total vehicle disinfection systems (Fig. 2b and 2c). Restriction of vehicle entry past a certain point on the property, physically separated from the birdhouse area, can also be effective where the vehicle does not need to enter the site (e.g. visitors and service personnel).
- Feed trucks are, by necessity, frequent visitors to poultry farms. The provision of remote feed delivery capacity (e.g. Fig. 3) can remove the need for the vehicle and its driver to enter the poultry holding area.
- Showers and changing facilities should be provided for high-value flocks and even flocks at lower levels on the hierarchy should provide clothing changing facilities. Where amenities for showering are provided, the design allowing staff to disrobe and enter the shower from one side and then emerge in a clean area where on-farm clothing and boots are provided is the most effective set up. Provision to achieve surface sanitation of articles being brought in can also be made, such as ultraviolet light cabinets, with openings from both sides of the change rooms. This can be used to sanitize necessary equipment and personal items (lunch boxes, etc.). It is my personal view that mobile phones should never be brought through onto the farm areas.

Figure 2 (a) Hand-operated vehicle disinfection, (b) automatic vehicle spray and (c) total vehicle spray and wheel wash.

Figure 3 Remote feed delivery system.

- Farms without showers should provide personal protective clothing for any visitor. On commercial broiler farms, this can be cheaply provided through disposable overalls, hairnets and provision of farm boots or the use of strong plastic boot covers.
- The entries to all houses should have a disinfectant footbath and a hand sanitizing device nearby which should be used on entry and exit by all persons. Footbaths need to be maintained and the sanitizer refreshed frequently. An even better approach is to provide dedicated boots for each house.
- Other livestock should not be allowed in the house vicinity, as these, and their faeces, can be sources of *Salmonella* contamination. Pets should also be restricted from the house area and should never be allowed inside a house.

3.1.5 Depopulation

In areas of dense poultry populations it can be advisable to attempt to coordinate flock ages and avoid large variations in broiler age within close proximity. This is essentially trying to make an area run in an 'all-in, all-out' mode. The benefit here is to avoid older flocks, especially those being depopulated, being in proximity to younger flocks. This can be a difficult situation to achieve but the benefits can be major, especially during local outbreaks of infectious disease, such as infectious laryngotracheitis in broiler flocks. This also requires considerable cooperation between different companies and organizations farming in the same locale.

3.1.6 Clean out, composting and re-use of litter

Preferably houses should go through a thorough batch cleaning and disinfection process before new chicks arrive. Difficulties in supply of clean litter materials, appropriate used litter storage and/or land application of used litter often hampers this, however, and it is necessary to re-use multi-batch litter. The latter may also provide some advantages, such as decreased environmental contamination with poultry manure and development of a more valuable fertilizer product. However, the re-use of litter should involve the proper composting (litter pasteurization) procedures for this to be safe and efficient. Good litter pasteurization requires proper litter heaping, monitoring of temperatures achieved during composting and sufficient time for the process to achieve the correct results of litter pasteurization and stabilization of nitrogen into microbial protein which is less volatile as ammonia.

In full clean-out processes, the general procedure is to remove all litter from houses and remove it from the vicinity. The house needs to be swept effectively to remove as much of the litter remnants as possible. The ceilings, walls, curtains, floors, feeders and drinker lines need to be washed – preferably with a heavy-duty detergent. All houses on the farm should reach this stage of preparation before a sanitizer is used (i.e. all litter removed and all houses washed). A sanitizer can then be applied to the clean facilities. An insecticide spray may also be applied after this procedure. The value of a detergent should not be underestimated and where time or economy is limited, use of a detergent should be given preference over a sanitizer. Then fresh clean litter material can be delivered. Some farms may also use a fumigation procedure (using formaldehyde gas, where allowed) once the litter is put in.

3.1.7 Free-range biosecurity issues

Exclusion of contact with wild birds is essential for truly effective biosecurity. Hence houses should be completely wild bird proof. Free-range operations obviously cannot completely exclude wild bird contact possibilities. The major concern here is possible direct or indirect contact with wild waterfowl or their faeces, with low but inherent risk of the introduction of low pathogenic AI virus. Free-range facilities should therefore do as much as possible to discourage the presence of waterfowl in the vicinity of the range areas. There should be no surface water sources (dams, creeks, rivers, drains, etc.) in the vicinity of the range areas. There should be no sources of feed or drinking water outside of the house. Any surface water used for the poultry (for drinking, cooling or irrigation of the range areas) must be effectively sanitized. Grass should be kept mowed short around and in the range areas. Any feed spills around silos or storage areas must be avoided and feed storage areas should be securely fenced off from access by the ranging poultry. Chickens that escape and are seen in the feed storage areas should be caught and culled. They should not be allowed to rejoin the main flock.

In most circumstances the range area is allowed a reasonably lengthy recovery period between batches of meat chickens. This would include the farm downtime while birds are absent and generally the first three weeks of the next flock's life before they are allowed to range. In environments such as Australia, with low humidity, high temperatures and high sunlight intensity, this length of time allows for reasonable natural disinfection of the range area. In more temperate and cooler climates this may need closer assessment with regard to the possibilities of pathogen survival and build up on protected range areas. Survival of large numbers of clostridial spores and coccidial oocysts may present issues depending on environmental conditions. Other possible pathogens could include helminth eggs and possibly histomonads, although these generally require longer lifespans than those afforded to modern meat chickens.

Good biosecurity is expensive. Like insurance, biosecurity is a necessary precautionary expenditure against incidents which we hope will never occur. Its expense often prompts broiler management to compromise and cut corners, particularly when there seems to be little disease operating, which may actually be an unrecognized result of good biosecurity. It is appropriate to consider biosecurity to be equivalent to a necessary insurance process, of similar importance as the financial insurance of physical assets of a business. Losses from disease incursion can be just as costly, and possibly more likely to occur than is, say fire damage to buildings.

In terms of relative effectiveness, isolation is the most effective biosecurity measure possible, and also the most expensive (farm location, access to staff and services). The next most effective measure is single-age (all-in, all-out) operation. Without these two factors, biosecurity relying solely on quarantine and hygiene measures is much more difficult. Reliance on the latter measures makes maintenance of biosecurity very hard and prone to failure.

3.2 Vaccination – breeders versus broilers

Typical vaccination programmes used in various countries for broiler breeders and broilers have been reviewed by Cserep[13]. Breeder vaccinations have the dual justifications of protection of the breeder flock from mortality and production losses and to ensure the provision of high maternally derived antibody for the progeny of the flock. The particular programme will vary depending on the diseases and their virulence and severity active in a particular location.

3.3 Egg-borne infections

Breeder flocks need to be free from or vaccinated against organisms known to be capable of transmission vertically; either within the fertile egg or on the eggshell. This will include the species-specific *Salmonella* serovars of poultry (Pullorum and Gallinarum), mycoplasmata (*M. gallisepticum, M. synoviae* and *M. meleagridis*), avian encephalomyelitis virus and chicken infectious anaemia virus in all regions and various other organisms that may be problematic in certain locations, such as reoviruses in many countries and fowl adenoviruses (e.g. types 8b and 11 in Australia). The paratyphoid serovars of *Salmonella* (particularly Enteritidis and Typhimurium) are also well capable of egg-borne transmission and freedom from these has proven much more difficult to achieve, but breeders supplying broiler chicks and poults should strive towards this goal. The status of the breeder flocks supplying day-old progeny should be known at least.

3.4 Management-related diseases and problems

More detail on management of particular diseases and conditions is provided further in this book, but the role of husbandry in all disease conditions must be recognized and accepted by the farmer in all situations. Although some conditions cannot be prevented at the final farm level, their severity and sequelae may be largely modified by good management conditions.

Brooding conditions can be major contributors to improved bird health. Examples of conditions which may modify an existing or introduced disease include increased severity of aspergillosis with high relative humidity (>80%) in the house, increased severity of infectious bronchitis with chilling and exacerbation of the latter prevalence of broiler ascites with slightly low brooding temperatures and prolonged photoperiods. Provision of correct brooding temperatures as advised by the breeding companies will aid in minimizing the deleterious outcomes of many situations. Correct temperatures of course include heating of the litter substrate to the correct temperature as well as achieving adequate ambient air temperatures, prior to the arrival of day-olds.

Attention to stocking density is crucial to good health, performance and welfare of the flock. Different localities, climates and housing systems will dictate the appropriate level. Overstocking is one of the major contributory factors to health issues in meat birds. A maximum density based on kilograms of bodyweight per square metre of floor space must be established and exceptional attention to not exceeding this must be practised. This is especially so where multiple thin outs are practised, as the first thin out of birds must occur in a timely manner to avoid the density target being over-reached. This can be a logistical challenge in some situations but deserves a strong commitment from management to ensure compliance. As a guide, maximum densities should not be allowed to exceed 34 kg/m^2 in tunnel-ventilated premises and possibly not more than 28 kg/m^2 in open-sided housing, depending on local conditions and house construction.

Nutrient deficiencies are uncommon in most commercial operations due to advanced feed formulation by experienced nutritionists but occasional problems can occur due to mistakes (e.g. premix or other ingredient left out, salt inclusion level incorrect), poor ingredient quality, toxicoses such as from mycotoxins or biogenic amines, contaminants, medication overdoses or *Salmonella* contamination.

Planned medications must be regularly assessed and challenged. Control of coccidiosis or necrotic enteritis may involve in-feed medicaments and adequate rotation or shuttle programmes need to be carefully considered and instituted to avoid or at least delay the development of resistance. Anticoccidial vaccines have undergone continued improvements in recent years with precocial strains providing good immunity without the subsequent need to medicate, but these still produce a mild intestinal infection that does temporarily reduce growth and may cause enough intestinal irritation to promote the occurrence of necrotic enteritis. But as the consumer pressure to remove feed medication from food-producing animals increases worldwide, the need to optimize the use and outcome of coccidial vaccines will continue to grow in importance.

3.4.1 Wet litter

Management of litter condition is another important consideration and this is gaining more prominence due to welfare accreditation programmes in some countries. Wet litter is a serious issue for bird health and is the major contributor to foot pad dermatitis development and cold, wet conditions contribute to severity of respiratory diseases. Contributory factors and causes of wet litter have been recently reviewed[11]. Broiler chickens drink about 1.8 litres of water per kg of feed consumed and 75% of this water ends up in the faeces. Ventilation is the only method available to remove moisture from the house. Ventilation is the major factor involved with litter moisture management and failure to ventilate adequately is the major cause of wet litter development, often closely linked with excessive stocking density. Hence there are more frequent problems seen in the cooler months as there is a compromise between air quality and temperature maintenance. Reduced ventilation leads to increased humidity and litter will become wet once relative humidity exceeds 70%.

Good management of drinkers (avoiding spillage, correct height and water pressure for the age of bird) is also of importance. Maintaining dry friable litter provides advantages: it has a higher gas transfer capacity and hence dries faster and is more resistant to compaction; it will be more alkaline and tend to promote aerobic conditions. Wet litter, however, has slower drying capacity, is easily compacted, promotes anaerobic conditions and tends towards acidity, with the increased risk of subsequent burning of the birds' feet and hocks.

Other factors of excessive moisture in the litter can include physical sources, such as water leaks or excessive condensation; bird health issues causing diarrhoea such as coccidiosis, necrotic enteritis, dysbacteriosis, infectious stunting syndrome, infectious bronchitis, colibacillosis and a myriad of viral infections (infectious bursal disease, Marek's disease, Newcastle disease (ND) and AI; and nutritional factors can be involved. Determining the initial cause of a wet litter problem can be difficult, particularly if the condition has existed for some time, but one approach is to determine the nature of the faeces predominating in the house[12].

Wet litter conditions where the majority of the faeces appear normal and well-formed is suggestive of presence of excessive water or failure to adequately remove moisture: hence ventilation may have been compromised or relative humidity may have been too high; there may have been lack of attention to litter aeration and cultivation or the litter depth or type of litter material may have been inadequate. A lighting regime that affects bird activity may also have contributed to wetter litter conditions as the lowered flock activity may reduce the flocks' ability to cultivate the litter.

Where faeces are observed to be watery (diuresis), some involvement of the kidneys could be suspected[12]. The possibility of renal damage can be investigated in areas including infectious bronchitis virus, presence of mycotoxins (e.g. ochratoxins) in feed and possibly avian nephritis virus infection. A possibility of an electrolyte imbalance (either through feed or from levels in drinking water) must also be considered.

The presence of orange/pink coloured mucous in faeces would indicate intestinal irritation[12] or disease, and examinations for the existence of dysbacteriosis or coccidiosis should be conducted. This can also occur if the birds have been off feed for a few hours.

Observation of undigested feed particles in faeces should alert one to the possibility of some level of infectious stunting syndrome, viral infections, oxidized fats in feed, coccidiosis or necrotic enteritis.

Nutrition-related causes of excessive diarrhoea can involve problems such as excessive electrolyte (sodium, potassium) inclusion in either feed or water (and this can be exacerbated by the use of electrolyte treatment via drinking water which may increase water intake and hence increase faecal moisture). Diarrhoea can also result from feed non-starch polysaccharides which are not controlled by the added enzyme in use. Excessive or imbalanced amino acid levels leading to high blood uric acid levels can stimulate water intake. High fat inclusions in rations may also result in diarrhoea or steatorrhoea. Unintentional presence of toxins (e.g. mycotoxins, rancid fats or biogenic amines) may also lead to excessively fluid droppings. Physical feed factors can also contribute to increase in faecal moisture, such as fine particle size which can cause intestinal irritation, whereas the inclusion of whole grain may reduce it.

3.5 Growth rate and associated problems

Modern commercial meat poultry strains have phenomenal growth rates; continuing improvements in this capacity is achieved each year as a result of genetic selection. Historically this also led to the inadvertent development of some health problems mainly involving the skeletal and cardio-respiratory systems. Leg weakness and ascites syndrome were the main examples. Rapid growth rates have been shown to impinge on bird health in these two problem areas[13]. Breeding companies have made substantial improvements in their breeds more recently and the industry has learnt ways of minimizing problems (e.g. by providing longer scotoperiods and heightened attention to brooding conditions). In some operations, there is a trend towards concentration on achieving higher seven-day weights as a major target. While this is a valid management performance target it should be viewed with some caution. Recent findings have shown that first week growth rates is increased in chickens which have hatched early and experienced a longer sojourn time in the hatcher which is associated with poorer leg health[13], and hence a focus here may need to be carefully considered.

3.6 Dead bird disposal

Removal of mortalities and culled birds is essential and must be done daily, usually the first action of the staff each day. Dead and sick birds are a source of contamination for other birds and their extended presence in a house may lead to problems with botulism. The mortalities must be disposed of so as to minimize the chances of further contamination of the farm itself and so as not to transmit disease to other farms. Environmental concerns often mean it is not possible to bury birds on site. Composting of dead birds is an effective method of dealing with them if done correctly. Incineration is also a good method if done correctly. Birds should not be burned in the open as feather material which may be infective may be spread by the convection currents evoked by the fire. Use of an enclosed incinerator is advisable. Otherwise the birds must be removed. Having a large freezer expedites this process so that mortality, if not excessive, may be stockpiled without massive decomposition and odour. Frozen birds can be removed at convenient intervals for disposal in public landfill or designated sites. Extreme care must be taken if vehicles for dead bird removal travel from farm to farm and thus pick up points for this process must be positioned to minimize the vehicle's entry to each farm site.

3.7 Hatching egg hygiene

The quality of day-old poultry is strongly reliant upon good husbandry and management of their parent breeder flocks. A major consideration here includes good hatching egg hygiene. Contamination of eggshells, especially soon after laying, is a major cause of reduced hatchability and of poor chick quality. This takes on greater significance as breeder flocks age and shell quality declines. Good hygiene practices of nest boxes and communal automatic nests on breeder farms are extremely important. Fumigation of eggs with formaldehyde gas has been a historic procedure that does improve the microbiological quality of eggs used for incubation. This practice has been withdrawn in many regions due to workplace safety requirements and hence a greater attention to nest and egg hygiene is required. Floor eggs should not be incubated, but under market conditions of egg shortages, the temptation is often provided to use lower quality eggs for incubation. If this does occur, poorer quality eggs should be separated and incubated separately and, if possible, kept separately as chicks in the field. More highly contaminated eggs will produce more day-olds with a propensity towards yolk sac infection and it may be beneficial to manage these chicks differently in the field as their first week mortality is usually increased. The use of antibiotics in such flocks is contentious. While this may appear to reduce first week mortality, quite often the effect is only to delay the problem into subsequent weeks and many such flocks experience increased incidences of problems such as femoral head necrosis (bacterial chondronecrosis and osteomyelitis). There is also the danger of provoking the development of multiple antibiotic resistance in bacterial populations in this situation. There are some indications that the use of probiotics in the first days of life for chicks may give some assistance with these problems[14].

3.8 Rodent and insect control

Rodent and vermin control measures form an essential part of quality assurance programmes for most meat chicken producing companies. Adequate positioning of bait stations which are refreshed frequently and monitored for rodent activity outside all houses are requirements of modern poultry production systems.

3.9 Multiple pickup for slaughter

The most desirable approach to house depopulation is to remove the entire flock in a single operation. Some regions, however, practice flock thin outs to reduce flock density at critical stages and meet bird size requirements for their markets while making more efficient use of facilities. This does introduce some disease spread risks as abattoir vehicles and equipment and catching crews move between farms, often during night-time. Where this is necessary it is important to limit the possible spread of contamination as much as possible by measures such as ensuring crews move from youngest to oldest farm and that attention to hygiene of catching equipment and vehicles is of a practically achievable level.

3.10 Free-range issues

The growth in free-range broiler management systems has been notable in several countries in recent years. This adds some complexity to farm management practices. Fortunately, with birds only released to range areas after three weeks of age, there is usually a lengthy downtime for the range areas, allowing for natural decontamination (exposure to sunlight, drying conditions, etc.) and time for regrowth of grass. Predator protection and discouraging of wild birds in the range areas assume more importance. Predation does not contribute markedly to bird losses directly but possible contact with wild animals and birds or their faeces does increase disease introduction risks. There should be no feed or drinking water supplied outside of the house to ensure that extra attractiveness to wild birds is avoided. Provision of shade is essential in range areas as chickens, being descended from jungle fowl, do not appreciate continual full sunlight exposure. If birds spend lengthy periods outside they are not consuming feed as contiguously as they would if fully confined and this can compromise coccidiosis control if it relies on in-feed anticoccidials. Market expectations in many countries also require the non-use of in-feed antibiotics, hence the prevalence of dysbacteriosis and occasionally necrotic enteritis may be higher. Alternative control measures for these intestinal conditions need to be developed for this method of production to be efficient or even sustainable in the long term.

4 Monitoring of poultry health and performance

Poultry companies record and store vast amounts of data on their flocks, primarily for production monitoring and reporting. Data covering flock location/identity, numbers of chicks placed, weekly average bodyweights, crude mortality and culls, feed conversion efficiency and pickup for slaughter information would be commonly kept. Histories of anticoccidial programmes and routine medications should also be documented. Some operations may also record breeder source identities, hatchery information, vaccination details, day-old chick weights and sex (if grown separately). Other more specific records are useful but seldom recorded routinely (e.g. disease-specific mortality or morbidity, lighting programme, farm visitor records, serviceman records of litter quality, ventilation assessments and problem notes). Laboratory records from breeder flocks are also available and taken to reveal the success of vaccination procedures and/or to evaluate the possible exposure to wild pathogens. Routine serological monitoring of broiler flocks, commonly for infectious bursal disease, infectious bronchitis and ND is also sometimes carried out. These resources are of immense value but are often underutilized. They are of use for the monitoring of performance and health. Often the computerized storage format is not easily amenable to formal analysis techniques and considerable editing is necessary to produce a database capable of straightforward statistical manipulation. The time needed to translate data into an efficiently analysable format is, however, completely worthwhile. A baseline for each parameter should be determined[15] for each operation. The knowledge of baseline data, seasonally adjusted, can be a useful technique for early detection of problems or for gaining an understanding of why performance may have improved. There is always a natural level of variability in field data, and assessment of performance must be considered within the framework of this variation to avoid invalid conclusions. A very instructive method of evaluating data on any of the parameters is to display them in a statistical control run chart[16]. Run charts plot data consecutively over time and confine the individual points between control lines calculated at three times the standard deviation (above and below the mean). As these are built up over time the value increases. These provide visual evaluation of trends occurring within an operation. Points plotted between the control lines represent natural variation. A single point outside the control lines would represent a significant outlier and an exception to the general process. Trends can become apparent from a successive run of points on one side of the mean line and can signal that performance is changing from a common source causation (either getting worse or better, depending on the direction). Identified changes can then be more thoroughly analysed against other known factors by standard epidemiological procedures. Any of the above parameters can be examined in this way.

There are systematic health assessment programmes in use in some operations where a sample of birds (often five birds per house) from different ages are collected and examined at post-mortem as a periodic snapshot of gross pathology. These often use a lesion scoring system to provide a record of a semi-quantitative assessment of disease signs occurring across the operation. An example would be the Elanco's Health Tracking System programme, but these carry on from suggested programmes developed by a university veterinary support group[17]. Such systems are extremely powerful and assist in the examination of health and production changes over long periods and provide inter-operation comparisons. These provide magnificent information but need to be more deeply examined, looking for correlations and associations within the recorded factors and to other farming factors to provide even more valuable outcomes.

4.1 Breeder monitoring

Commercial broiler health is heavily dependent upon the health status of their parent flock. Parent breeder flocks may be a source of vertically transmitted organisms and also a source of essential maternally derived antibodies for the young broiler. The latter provision is responsible for an enormous amount of protection for the young chick for up to three weeks of its life, which now represents 30% or more of the broiler's entire lifespan. Breeder flocks are extensively vaccinated with this outcome in mind. The particular diseases where maternal antibody contributes to prevention or management will vary between geographic locations but usually include infectious bursal disease, chicken anaemia virus, fowl adenovirus (e.g. types 8b and 11 in Australia), reoviruses and avian encephalomyelitis. In some situations, maternal antibody can interfere with the quality of vaccinations given to broilers (e.g. Marek's disease and infectious bronchitis) and is also of critical importance in countries where broilers themselves receive live infectious bursal disease vaccines. Serological monitoring of the breeder flocks is imperative to ensure that an adequate transmission of maternal antibody occurs for the diseases endemic to their locality.

Many breeder operations are either free of mycoplasmata or vaccinated against them, and the removal of this source of infection has improved broiler health dramatically. Paratyphoid *Salmonella* serovars, however, remain particular organisms of concern for potential transmission through the human food chain and hence will be monitored by environmental swabbing and culture in breeder flocks.

Hatcheries should provide the broiler farms with health status information on the source flock(s) for their chicks so that appropriate managements can be instigated or varied as necessary.

5 Management of sick birds

5.1 Diagnosis

Arriving at a correct diagnosis is obviously essential. The processes described in Section 2 should be observed. Clinical signs and gross pathology are often sufficient to arrive at a preliminary diagnosis in many conditions. Laboratory and further methods are used frequently and should be applied for confirmation and often for more accurate identification of the strains of the infective organisms involved. The latter investigations are becoming increasingly important in the understanding of the epidemiology of the disease in question and may be very important in the development of future control measures. Appropriate tests for various diseases have been reviewed[18]. In any poultry disease occurrence where there is a relatively large increase in mortality or an unexplained drop in egg production, it is imperative that the possible presence of an emergency disease be excluded. The clinical and gross pathological signs shown in ND and highly pathogenic AI are not sufficiently specific enough to be ruled out. In all such cases, where it is possible, tracheal and cloacal swabs, collected in a viral transport medium, should be submitted to a government laboratory capable of rapidly identifying the presence of these viruses.

Individual sick, lame or injured birds should be euthanized humanely as soon as they are observed. Dead or culled birds should be removed from the house promptly. Setting up 'hospital pens' in large commercial operations is generally inadvisable as very ill birds seldom recover and they act as a source of continued infection. Culling and removal of these animals is appropriate.

5.2 Mass medication methods

The sheer size of modern poultry flocks makes administration of treatment to individual animals logistically and economically unfeasible. Rapid medication of a large flock with a therapeutic product will require administration via drinking water. Poultry houses should each have a water-holding tank that can be used for medication purposes. Some live vaccines are also administered in this way. Dose rate calculation needs to be carefully considered and withdrawal times recorded and observed. Time of administration is important, as it is desirable that the medication is consumed promptly. If administered late in the afternoon, a proportion of the medicated water may sit unconsumed overnight and could be compromised by temperature or time. Administration of vaccines via drinking water requires considerable attention to detail if a successful result is to be obtained. This process usually requires withdrawal of water for an hour or more to ensure the birds are desiring to access the drinkers promptly. Water stabilizers or chlorine inhibitors are usually mixed into the water to receive the vaccine (e.g. skim milk powder and proprietary stabilizers) prior to vaccine suspension. Thorough mixing of the diluted vaccine into the required amount of water for about 2–3 hours of supply is essential. The water containing the vaccine should then be flushed through the drinking lines to ensure an immediate dose is available to the birds and the house should be walked by staff to encourage the birds to move towards the drinkers.

Automatic proportioning devices are also available for adding medication directly into a flowing water line. These may work well for chemical medications but it is difficult to adequately apply an in-water vaccine through this method. Close attention needs to be paid to ensuring that correct dose is achieved across the appropriate time frame required. Where water stabilization of vaccines is required an adequate amount of the stabilizer needed for the entire volume of water to be medicated must be supplied through the proportioner, bearing in mind that it will be difficult to achieve the time of water contact needed to achieve stabilization in some cases.

In-feed medication may be used as a follow-up to water administration of medicines or for longer-term medication. Detailed attention is needed to ensure proper withdrawal times are observed with in-feed medications, however.

5.3 Responsible use of antibiotics

Concerns over the development of antibiotic resistance which may eventually be transferred to the human population are increasing worldwide. Chemotherapeutic products should be under strict veterinary control in all cases. There should be a strong justification for their use to be considered. In commercial chickens, the eradication or control of *Mycoplasma* in breeding flocks, improved biosecurity and better control of immunosuppressive diseases (Marek's disease, infectious bursal disease, chicken anaemia virus) have largely removed the need for therapeutic medication of broiler flocks. Medication for problems such as yolk sac infection is generally unrewarding and should only be used where there are no reasonable alternatives (and management and other practices as described above should take precedence). Withdrawal times of medications need to be strictly observed to avoid meat residues.

The use of in-feed low-level prophylactic antibiotics ('antibiotic growth promoters') has been an historical feature of intensive animal industries. Because of possible antibiotic resistance development concerns, this practice has been banned or is under review in many countries. In many instances, the removal of this prophylaxis resulted in increased clinical disease, especially necrotic enteritis and dysbacteriosis, and actually led to an increase in therapeutic drug use. While arguments rage about the pros and cons of the practice, responsible use of antibiotics is important. Where in-feed prophylactic antibiotic use is allowed, there are obvious measures regarding the choice of such products that could be instituted to avoid or minimize the undesirable outcomes considered. Choice of antibiotics used in this way should be limited to those which are not orally absorbed by the chicken (i.e. only have gut activity), have a limited spectrum of antibacterial activity (e.g. Gram-positive spectrum) and are not chemically related to any antibiotic used in human medicine.

5.4 Communication with the grower/serviceman/manager

Rapid communication between grower and the management of the operation is imperative. The service personnel are usually the first link to a veterinarian and their experience is invaluable in first-line appreciation of the development of a problem that may require veterinary attention. Similarly, it is important for the veterinarian to communicate findings, laboratory results, treatment recommendations and regulatory requirements through the management structure back to the grower. In this way everyone involved is aware of the progress of the situation and mistakes and misunderstandings can be minimized.

5.5 Economic judgement

Final decisions about any health matter including treatment rests with the owners of the birds. In some situations the overwhelming costs may be the deciding factor for the operation in any treatment option. Having said that it is the management's responsibility to make the welfare of the animals and the safety of any food produced from them as the primary consideration, these must override any economic consequences.

6 Emergency disease occurrence

In many countries the outbreak of serious poultry diseases will result in government authority intervention and attempts to eradicate the disease. In commercial meat chicken flocks, the two diseases likely to trigger a disease emergency response are AI[19] and ND[20]. The difficulties with initial recognition and diagnosis of these diseases are covered elsewhere but clinical signs and post-mortem pathology of these diseases may be difficult to distinguish from other serious respiratory diseases including fowl cholera, infectious laryngotracheitis, chlamydophilosis, acute toxicities or management malfunction (e.g. overheating and smothering).

It is in the interests of the local industry, the region and the nation to expedite the detection of outbreaks of emergency diseases as soon as possible. Outbreaks of such diseases need to be contained and eradicated promptly. Whenever confronted with a clinical picture in broilers involving high mortality with respiratory or nervous signs even if other conditions are strongly suspected, steps should be taken to quickly exclude and/ or confirm highly pathogenic AI and ND. Swabs collected from trachea and cloaca from a meaningful number of freshly dead or clinically ill birds should be submitted without delay to a government laboratory, preferably transported in a viral transport media. PCR tests are capable of detecting HPAIV and velogenic NDV within 12 hours of arrival at the laboratory.

6.1 Mass destruction techniques

Control of emergency diseases such as AI and ND in many countries is reliant upon eradication which entails the need for rapid mass destruction of poultry flocks. Flocks may also need to be destroyed where there has been a market failure or malfunction, especially so for large broilers that cannot be normally slaughtered and are approaching critical body mass or growing too large in a house, severely compromising their welfare. Methods of killing large numbers of poultry have been reviewed[21,22]. Euthanasia methods that are deemed to be acceptable for poultry include occipitocervical dislocation, decapitation and gassing[21]. CO_2 is the most frequent gas used in mass destruction and this requires the confinement of birds to a contained structure, such as a waste skip bin[23] or a purpose-designed killing cart[21]. The birds need to be caught and manually placed into these devices and the process can cause stress, not only to the birds but also to the people involved. Such procedures are never pleasant and can be a considerable source of distress for owners of the birds and staff associated with the farm. Mass destruction processes are desired to be rapid to interrupt the possible spread of infection. With this end point in mind, welfare needs to be compromised to some extent to enable the destruction. It is important to carry out this procedure in the most humane manner possible in the situation. More recently, fire-suppressing foam has been employed as a method of destroying a large number of birds *in situ* in the house. The birds do not drown with this procedure. The foam kills the birds by asphyxiation within a few minutes of being covered above their heads. The process is not immediate: movement can be appreciated underneath the foam for several minutes. This procedure is rapid and efficient but does not cause immediate death on application. Under the OIE *Terrestrial* Code, the principles needed to be considered for humane destruction refer to the induction of unconsciousness to be under 'the least aversive conditions as possible and should not cause avoidable anxiety, distress or suffering in animals'[21]. In this sense, use of foam does overcome the stressful process of manually herding, catching and carrying of birds to a gassing chamber or the distressing practice of mass neck dislocation within a farm situation. Care needs to be taken in applying the foam to large groups as an advancing foam front may cause birds to panic and flee, leading to packing up of birds. Dividing the flock into smaller groups with portable solid fences may assist in this regard (Fig. 4). The foaming technique is usable where the birds are on the floor but is difficult if the floor is slatted and will not work well in a cage facility.

6.2 Disposal of mass mortality

Masses of potentially contaminated killed birds should be disposed of promptly in a manner that minimizes the chance of further spread of the infection. Effective methods are burial, incineration, rendering and composting (the latter can be completed on site (Fig. 5), avoiding the need to transport dangerous material)[24].

Figure 4 Penned section within a house filled with firefighting foam for mass destruction of poultry.

Figure 5 (a) Composting of mass mortality *in situ*; (b) monitoring of compost temperatures.

7 Future trends and conclusion

Challenges that will remain as issues for developed nations into the near future will include areas of food safety, the move away from the use of antibiotics in production, animal welfare, environmental sustainability and diminishing worldwide resources. Continuing emphasis on endemic diseases and emergency diseases, AI in particular, will also be important. In the developing world maintenance of poultry farming is a priority with local industries experiencing sustainability difficulties[24].

The rapidly increasing world human population will continue to drive the need for food production and poultry meat appears as the most sustainable industry that would be able to meet the growing demand for protein. Poultry production is the most efficient converter of food sources to high-quality protein, produces no liquid waste, uses less water and produces the least amount of greenhouse gas emissions of any of the livestock systems.

7.1 Food safety

High-quality and safe food is the goal of the industry and is demanded more and more by safe food authorities and consumer groups. In some countries the standards required by regulatory authorities extend beyond meat processing right back through the production chain. Enhanced traceability of product has become essential in some jurisdictions and can only be seen to be an increasing requirement in the future[25]. Standardized control procedures such as hazard analysis and critical control point will continue to be important and will be audited even more resolutely. Even though contamination of the final product can be effectively controlled during the slaughter and processing system there is clear desire to decrease the amount of human pathogenic bacteria which come from the farming sector. The major pathogens of interest with poultry meat being *Salmonella* serovars (particularly serovars Enteritidis and Typhimurium) and *Campylobacter* species. Efforts to control these more effectively at the farming level will continue to be driven in the foreseeable future.

7.2 Animal welfare

Consumers, animal welfare interest organizations, major retailers and fast food chains are placing ever-increasing concern on how animals are raised in agriculture. A poor welfare image to the public has been a burden on the poultry industry for many years and activist groups continue to apply pressure in this aspect. Meat chicken growers do have a serious view on the well-being of their birds and genuinely wish to maximize their welfare while still being able to farm sustainably. The clash of views has been volatile in the past and will continue to follow a confrontational path unless the industry takes a proactive role in continually improving and being seen to be improving their systems. The huge increase in the popularity of alternative systems in some countries, notably the move to free-range production systems, has been facilitated and in some cases forced through the increased public interest in animal welfare. As we have noted, moving to alternative production systems is not without the risks of reduced productivity and increased disease prevalence and the industry needs to continually research and evaluate improved methods to actually enhance welfare once moving into alternative systems. This will be a continued area of growth in research expenditure and in communication between industry, government regulators and consumer groups. The introduction of universally accepted standards is becoming an important step towards this end goal.[25,26]

7.3 Environmental sustainability

Chicken meat production is the most environmentally friendly of the livestock meat industries, using less land, more efficient conversion of feed to protein, less water use per kg of meat produced (including that for the feed needed) and lower waste production and the lowest greenhouse gas production.[27] Chicken meat appears as the most likely sustainable source of high-quality protein to support the predicted rise in global human population. Even so the reduction of the industry's environmental footprint remains a challenging goal for sustainability. Issues involving the proximity of urban living to poultry farms continue to cause concern and problems for planning authorities. The main issues here are odour production, visual pollution and noise associated with vehicle movements during collection and transport of birds for slaughter. Better future planning regulations may help decrease these issues in future but they will remain as a problem area for some time yet, particularly as the urban spread encroaches into traditionally rural areas on city fringes. This will encourage the need to move and re-establish poultry enterprises in new regions.

7.4 Diminishing resources

As the human population grows, all resources will come under ever-increasing pressure. Adoption of newer technologies will be essential if the chicken meat industry worldwide is to meet the protein needs of a burgeoning consumption base. One of the major limitations will be sources of phosphorus. The industry has moved to improved utilization of phosphorus sources with the introduction of phytases to make plant phosphorus sources frequently available. Continued improvements will be necessary going forward. Another major challenge is emerging as competition for grains from the ethanol production industry. As the world searches for renewable energy sources, a threat to available food sources emerges and this will need to be carefully balanced into the next two decades.

7.5 Challenges for poultry in the developing world

Perhaps the largest challenge to the sustainability of chicken meat production in developing countries comes from the importation of lower cost meat from other high-producing countries, providing competition with which the local industry cannot compete. The cost of production in developing countries is a major obstacle, the major element being feed costs, which can only be seen to increase in the near future. Returns on their produce is also limited often due to lack of market access[28]. Interest rates and access to finance is also more difficult in the developing world[24]. Access to products to support efficient production, particularly vaccines, in some countries is limited[29]. Inadequate control of diseases and lack of implementation of biosecurity remain as areas where vast improvements in productivity could be made in this region of the world. Water quality is an issue for animal health as much as it is for human health, and its improvement offers massive increases in the standard of living for many regions.

8 Where to look for further information

Sources of information on health investigation include:

Martin, S. W., A. H. Meek and P. Willeberg (1987), *Veterinary Epidemiology: Principles and Methods*. Iowa State University Press, Ames, Iowa, USA.
Pattison, M., P. F. McMullin, J. M. Bradbury and D. J. Alexander (eds) (2008), *Poultry Diseases*, Sixth Edition. Elsevier, Edinburgh, UK.
Saif, Y. M. (ed.) (2008), *Diseases of Poultry*, Twelfth Edition. Blackwell Publishing, Oxford, UK.
Thrusfield, M. (ed.) (2017), *Veterinary Epidemiology*, Third Edition. Blackwell Publishing, Oxford, UK.

9 References

1   Bermudez, A. J. 2008. Principles of disease prevention: Diagnosis and control introduction. In *Diseases of Poultry*, 12th Ed., Y. M. Saif (Ed.), pp. 3–46. Blackwell Publishing Ltd.: Oxford, UK.

2   Alcorn, M. J. 2008. How to carry out a field investigation. In *Poultry Diseases*, 6th Ed., M. Pattison, P. F. McMullin, J. M. Bradbury and D. J. Alexander (Eds), pp. 14–38. Elsevier: Edinburgh, UK.

3   Lessard, P. 1988. The characterization of disease outbreaks. *Vet. Clin. North Am. Food Anim. Pract.*, 4:17–31.

4   Hueston, W. D. 1988. Evaluating risk factors in disease outbreaks. *Vet. Clin. North Am. Food Anim. Pract.*, 4:79–96.

5   Willeberg, P. 1979. The analysis and interpretation of epidemiological data. *Proceedings of the Second International Symposium of Veterinary Epidemiology and Economics*, pp. 185–90. Canberra, Australia.

6   Hancock, D. D. and Wikse, S. E. 1988. Investigation planning and data gathering. *Vet. Clin. North. Am. Food Anim. Pract.*, 4:1–15.

7   Groves, P. J. 1999. The ascites syndrome in broiler chickens: identification of epidemiological risk factors associated with the syndrome and examination of their nature. PhD Thesis, The University of Sydney, Sydney, Australia.

8   Blood, D. C. and Studdert, V. P. 1999. In *Saunders Comprehensive Veterinary Dictionary*, 2nd Ed., p. 132. WB Saunders: London, UK.

9   Groves, P. J., Walkden-Brown, S. W., Islam, A. F. M. F., Reynolds, P. S., King, M. L. and Sharpe, S. M. 2008. An epidemiological survey of MDV in Australian broiler flocks. *Proceedings 8th International Marek's Disease Symposium*, p. 26. James Cook University: Townsville, Queensland, Australia.

10   Cavanagh, D. and Naqi, S. A. 2003. Infectious Bronchitis. In *Diseases of Poultry*, 11th Ed.,Y. M. Saif (Ed), p. 107. Iowa State Press: Ames, Iowa, USA.

11   Dunlop, M. W., Moss, A. F., Groves, P. J., Wilkinson, S. J., Stuetz, R. M. and Selle, P. H. 2016. The multidimensional causal factors of 'wet litter' in chicken-meat production. *Sci. Total Environ.*, 562:766–76.

12   LaVorgna, D. and Vancraeynest, D. 2104. Evaluating wet droppings as a useful guide to bird health. *Int. Poult. Prod.*, 22:11–13.

13   Julian, R. J. 2005. Production and growth related disorders and other metabolic diseases of poultry–a review. *Vet. J.*, 169:350–69.

14   Wideman, R. 2016. Probiotics as an alternative to antibiotics for treating lameness due to bacterial infections in broilers. *Proceedings APSS*, 27:200–5. University of Sydney: Sydney, Australia.

15   Keirs, R. W. 1993. Flock health monitoring programs. *Proceedings XIII Latin American Poultry Congress*, pp. 29–36. Santo Domingo, Dominican Republic, October.

16   Tague, N. R. 2005. *The Quality Toolbox*, 2nd Ed., pp. 155–8. Milwaukie, USA: ASQ Society for Quality. Available at http://asq.org/learn-about-quality/data-collection-analysis-tools/overview/control-chart.html.

17   Keirs, R. W., Magee, D., Purchase, H. G., Underwood, R., Boyle, C. R. and Freund, J. 1991. A new system for broiler flock-health monitoiring. *Prevent. Vet. Med.*, 11:95–103.

18   Morrow, C. 2008. Laboratory investigation to support health programmes and disease diagnosis. In *Poultry Diseases*, 6th Ed., M. Pattison, P. F. McMullin, J. M. Bradbury, D. J. Alexander (Eds), pp. 39–47. Elsevier: Edinburgh, UK.

19   Animal Health Australia. 2011. Disease strategy: Avian influenza (version 3.4). *Australian Veterinary Emergency Plan (AUSVETPLAN)*, Edition 3, Primary Industries Ministerial Council: Canberra, ACT, Australia. Available at http://www.animalhealthaustralia.com.au Validated 18 August 2016.

20   Animal Health Australia. 2014. Disease strategy: Newcastle disease (version 3.3). *Australian Veterinary Emergency Plan (AUSVETPLAN), Edition 3*, Agriculture Ministers' Forum: Canberra, ACT, Australia. Available at http://www.animalhealthaustralia.com.au Validated 18 August 2016.

21   Thornber, P. M., Rubira, R. J. and Styles, D. K. 2014. Humane killing of animals for disease control purposes. *Rev. Sci. Tech. Off. Int. Epiz.* 33:303–10.

22   Tahir, R., Anjum, A. A., Muhammad, K., Rasool, A. and Khan, F. 2015. Avian influenza and its mass depopulation strategies in infected poultry birds. *Taiwan Vet. J.*, 41:51–7.

23 Animal Health Australia. 2015. Operational Procedures Manual: Destruction of animals (version 3.2). *Australian Veterinary Emergency Plan (AUSVETPLAN)*, Edition 3, Agricultural Ministers' Forum: Canberra, ACT, Australia. Available at http://www.animalhealthaustralia.com.au/ Validated 18 August 2016.

24 Kusi, L. Y., Agbeblewu, S., Anim, I. K. and Nyarku, K. M. 2015. The challenges and prospects of the commercial poultry industry in Ghana: a synthesis of literature. *Int. J. Manag. Sci.*, 5:476–89.

25 Penz-Jr., A. M. and Bruno, D. G, 2011. Challenges facing the global poultry industry until 2020. *Proceedings of the Australian Poultry Science Symposium*, 22:49–55, Sydney, Australia.

26 Australian Chicken Meat Federation, 2013. Bird Welfare. Available at http://www.chicken.org.au/page.php?id=44 Accessed 14 January 2017.

27 de Vries, M. and de Boer, I. M. J. 2010. Comparing environmental impacts for livestock products: a review of life cycle assessments. *Livest. Sci.*, 128:1–11. doi:10.1016/j.livsci.2009.11.007.

28 Karthikeyan, S. 2012. Problems faced by small scale poultry farmers in developing countries. *The Poultry Site*. Available at http://thepoultryguide.com/problems-of-small-scale-poultry-farmers/Accessed 14 January 2017.

29 Anonymous, 2008. Plenty of poultry farming challenges. *Poultry News*, The Poultry Site available at http://www.thepoultrysite.com/poultrynews/15627/plenty-of-poultry-farming-challenges/ Accessed 14/01/2017.

Chapter 4

Improving biosecurity in poultry flocks

Jean-Pierre Vaillancourt and Manon Racicot, Université de Montréal, Canada; and Mattias Delpont, École Nationale Vétérinaire de Toulouse, France

1 Introduction

Biosecurity is any actions or health plans designed to protect a population against infectious and transmissible agents (Toma et al., 1999). They are based on principles that have been known for centuries. Sanitation measures are detailed in *Deuteronomy*, written about 2600 years ago (Ojewole, 2011); most biosecurity measures found in modern poultry farms appear like analogies of measures prescribed to lepers during the mass of separation in the Middle Ages (Brody, 1974). Yet, we are still struggling today to get people to perform simple tasks that we know work for preventing contagious diseases on poultry farms. For example, in 2011, Racicot et al. identified 44 different errors when getting in and out of poultry barns. Most errors had to do with area delimitation (clean versus contaminated areas of the entrance or anteroom of a barn), boots, and handwashing.

Biosecurity on poultry farms has been categorized in different ways (internal, external; operational, structural; bioexclusion, biomanagement, biocontainment, etc.). Ultimately, all on-farm biosecurity measures have to do with two principles: reducing sources of

http://dx.doi.org/10.19103/AS.2022.0104.05

contamination and separating these sources from healthy flocks. In modern poultry production, a third principle linked to the latter is emerging: the need for a regional approach to biosecurity based on communication within the poultry industry and the organization of movement of people, birds, material, and equipment within any given region. Indeed, the intensification of poultry production has created an environment that may promote the spread of contagious diseases. Any poultry activity involves inherent infectious disease transmission risks, and the amplitude of these risks increases with regional poultry density (Fernandez et al., 1994).

However, no matter how we define or partition biosecurity measures, the recurring challenge is the consistent implementation of these measures by all farm personnel and visitors. Achieving high compliance with biosecurity measures is, therefore, a constant issue that needs addressing in terms of research priority.

2 Reducing sources of contamination: cleaning and disinfection of poultry barns

The decontamination of poultry barns is a multistep process. The first step is the dry removal of organic material. It consists of removing with physical force as much visible organic material as possible: litter, feed, dust, etc. The design of the building and its equipment can significantly affect cleaning (Cerf et al., 2010). For example, the floor of an anteroom with ground equipment will be more difficult to clean on a regular basis compared to one where most equipment is suspended.

The second step is the application of a detergent followed by stripping and drying: some organic materials or mineral deposits are impossible or very difficult to remove using only physical force (Step 1). Step 2 is, therefore, the step in which this material is removed using a chemical process, such as a detergent. It makes it possible to reduce the bacterial concentration of a contaminated surface by a factor of 10-1000 compared to a wash with only clean water (Course et al., 2021). Using a detergent also helps reduce the total amount of water used. Several chemicals influence the effectiveness of detergents: surfactants, wetting agents, dispersants, pH adjusters, and sequestering agents. However, two main categories are recognized: acid and alkaline detergents. A rule of thumb is that acid detergents remove mineral deposits, while alkaline detergents remove organic waste not removed by dry cleaning (i.e. dried, greasy, or sticky material). The choice of detergent, therefore, depends on the type of material present at the start and the water used. If the cleaning water used is hard (>120 ppm), more frequent use of an acid detergent is necessary (Watkins and Venne, 2015).

When applying the detergent, there are several rules to follow. Firstly, the contact time: the use of detergent in foamy form with the help of specialized equipment is preferred because this makes it possible to optimize the contact time, which should be 10-20 min (Springthorpe, 2000). Second, the temperature of the cleaning water. If hot water is available, it should be used. Water that is too cold slows down chemical reactions. Water at 40°C is best (Böhm, 1998). Third, the flow and pressure. The flow rate and quantity of detergent are greater when applying the detergent, while the pressure is greater during stripping, which is the action following the application of the detergent. When applying the detergent, it is recommended to use a flow rate of 10-20 L/min at low pressure (300-500 psi) and to use approximately 250-500 mL of solution per square meter (Blondel et al., 2018).

If the application of the detergent takes more than 20 min, then the stripping should be done alternately with the application of the detergent. Stripping is the action of removing the rest of the organic debris that the detergent has made accessible. Using clean high-pressure water (1000–3000 psi) at a 45° angle works well for this step, as it adds mechanical force (Blondel et al., 2018). However, it is important to avoid damaging the surfaces; such pressure can damage wooden structures, creating crevices making future decontamination more difficult (Blondel et al., 2018).

Finally, it is important to let the areas dry before proceeding to the next step. Omitting this step may reduce the effectiveness of disinfection in two ways: by dilution and by inactivation on contact with organic material. The effectiveness of a disinfectant is related to its concentration when applied. The presence of water at the surface of what needs to be disinfected dilutes the disinfectant and reduces its effectiveness (Böhm, 1998).

The third step is disinfection, followed by drying. This step should only take place when all organic material has been removed and the areas are dry (Langsrud et al., 2003; Payne et al., 2005). It is important to visually monitor the area that has been cleaned before disinfection (Fig. 1). A nonpermanent marker such as a bright color aerosol hair spray or a tape can be used to identify where cleaning is not satisfactory (presence of organic material); this way, it is not necessary to completely redo the cleaning process for the entire barn. This approach may show over time areas that require special attention.

Disinfection can be carried out by vaporization (liquid or foam) and/or by fumigation/misting. Regardless of the method and product used, it is important to follow the manufacturer's recommendations for concentration, contact time, product mixture, etc. (Payne et al., 2005). On average, 150–300 mL/m² is applied (Blondel et al., 2018).

The choice of disinfectant depends on several factors: the surface to be disinfected, the pathogens targeted, the hardness and the pH of the washing water, the cost, etc. Generally, the commercial disinfectants available for livestock buildings consist of

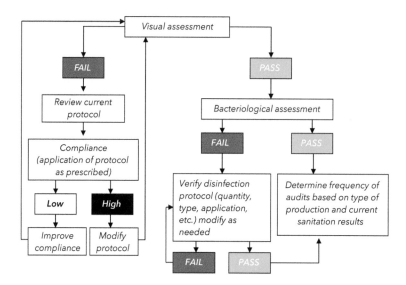

Figure 1 Pathway to assess an on-farm cleaning protocol.

active ingredients from the following main families: halogenated derivatives, aldehydes, quaternary ammoniums, phenols, and peroxide. Drying the building after the application of the disinfectant is essential, as it prevents the multiplication of the remaining bacteria (Davies and Wray, 1995), and this desiccation period may contribute to reducing further the number of living organisms (Böhm, 1998). In some countries, heating the house (40°C) for 3-4 days has been useful (e.g. preventing the spread of *Mycoplasma gallisepticum* in North Carolina, USA). Finally, a microbiological assessment of the disinfection process may be performed as part of the monitoring (Fig. 1).

3 Reducing sources of contamination: equipment and vehicles

Vehicles and equipment may act as a mechanical vector. Dee et al. (2002) demonstrated that snow contaminated with the Porcine Reproductive and Respiratory Syndrome (PRRS) virus and applied to the undercarriage of a car can contaminate the environment for several kilometers. The inadequate management of vehicles, including where they park on a production site (too close to the barn entrance), has been associated with a greater risk of disease transmission between farms (Guinat et al., 2020). Rose et al. (2000) demonstrated an association between the surface covered by vehicles on a broiler chicken production site and the probability of isolating *Salmonella*. Snow et al. (2010) also identified a link between on-farm transportation and *Salmonella* in egg layer farms. Vehicle-sharing practices between poultry production sites are also associated with the spread of infectious laryngotracheitis and avian influenza (Nishiguchi et al., 2007; Volkova et al., 2012).

Wheel baths are assumed to be useful (Pinto and Urcelay, 2003; Zhang et al., 2013), but there is little evidence that they work. In fact, several arguments have been presented to support their lack of effectiveness. Briefly, wheels rotating at high speed at a high temperature over dry surfaces are not likely to harbor pathogens; even if they have mud on them, passing quickly in a wheel bath does not allow the disinfectant, assuming the solution is fresh, to reach the pathogens; finally, the contact time is too limited for disinfectants to reduce wheel contamination. Pathogens in vehicles are more likely to be present in the loading area and in the cab of the vehicle (Chaber and Saegerman, 2017).

The recommendations made for decontaminating poultry barns essentially apply to the decontamination of vehicles and equipment. When possible, they should be exposed to the sun, which helps the drying process as well as decontamination. Indeed, the ultraviolet radiation in sunlight works as a natural disinfectant (Kuney and Jeffrey, 2002).

4 Reducing sources of contamination: water and feed hygiene

4.1 Water hygiene

Water quality and sanitation would require its own chapter. Briefly, when using chlorine, it is best to assess the oxidation reduction potential (ORP) of water. It measures in millivolts the oxidizing potential of free chlorine residual. A strong oxidizer will effectively destroy microbes. An ORP value of 650 millivolts or greater is required. The goal should be to maintain 2-5 ppm of free chlorine in the water supply. Free chlorine is the residual available for sanitation (Watkins and Venne, 2015). For removing scale and biofilm, it is best to drop the water pH below 5 but not below 4 to prevent equipment damage. A bleach solution might be effective in removing biofilm, but it may also damage equipment. Several disinfectants are available to clean water lines, but some of the most effective and safe products for drinker systems are concentrated, stabilized hydrogen peroxides. Biofilms or any growth of bacteria, molds, and fungus in drinker systems can only be removed with cleaners that contain sanitizers (Watkins and Venne, 2015).

4.2 Feed hygiene and delivery

Basic feed mill biosecurity focuses on tight and closed silos to prevent rodent access, a continuous integrated pest management program, and sanitation procedures, mainly for silos that may have been contaminated with pathogens of importance to poultry. The main one, in terms of public health, is *Salmonella* (Jones, 2011). Li et al. (2012) reported that 8.8% of ingredients of animal origin obtained from three feed mills were contaminated with *Salmonella*, but dust samples had a higher contamination rate of 18.5%. Poultry offal meal and feather meal should be considered high-risk ingredients (Butcher and Miles, 1995). Several chemical options have been tested for the treatment of feeds to control *Salmonella*. Formaldehyde is one of the more frequently used chemicals under commercial conditions (Ricke et al., 2019). Cochrane (2016) provide detailed information on feed mill biosecurity plans.

Basic biosecurity for feed delivery starts with a check with dispatch for routing instructions to avoid passing by potentially diseased production sites or going from one to a disease-free operation. The driver of a clean feed truck must wear clean clothes and have plastic boots, hand cleaning solution, and fly spray available when entering a farm; the driver must never enter barns past the contaminated zone, in other words, must not get on the clean side of the anteroom (Jones, 2011). After all runs are completed, the driver goes directly back to the feed truck area. Trucks must be washed completely and disinfected weekly, unless the last site visited has birds infected with a disease of significance to the industry. In this case, truck decontamination must be performed after this last delivery. Breeder farms are normally visited first before meat bird operations. Delivery to diseased or quarantined farms should be done as the last load of the day (Anon., 2018).

5 Reducing sources of contamination: insect, mite and rodent pests, wild birds and pets

An often-overlooked vector, but with a significant risk of disease transmission, is the common house fly, *Musca domestica*. In 1975, more than 100 different pathogens that could be carried by flies and more than 65 diseases transmitted to humans and animals, such as salmonellosis, had already been identified. They can travel more than 30 km, but the majority of flies are confined to an area of 3 km^2. They can move quickly over long distances: 1–1.5 km in 24 h, 3–5 km in 48 h, and 6 km in 72 h (Schoof, 1959). Disease transmission can therefore occur between neighboring farms. Flies can also pose a public health problem. In 2009, researchers isolated enterococci and staphylococci in the digestive tract and on the exterior surfaces of flies caught near poultry farms. These bacteria were resistant to antibiotics of importance in human medicine (Graham et al., 2009). A study showed that 8.2% of the flies tested within 50 m of a farm were positive for *Campylobacter jejuni* (Hald et al., 2004). A study in Japan on highly pathogenic avian influenza H5N1 showed that within a kilometer of the farm, 30% of the flies were contaminated with the same virus as the poultry; at 2 km, the percentage decreased to 10% (Sawabe et al., 2006).

Besides flies, mealworms or darkling beetles (*Alphitobius diaperinus*) are an important concern in disease control. They can transport and transmit microorganisms such as *Campylobacter*, *Salmonella*, infectious bursal disease virus, reovirus, and Eimeria. The mealworm is one of the most abundant insects in poultry farm litter, reaching levels as high as 1000 mealworms per square meter. The risk of disease dispersal can therefore be considerable when spreading litter. In addition, chickens, wild animals, and rodents feed on these insects and can thus infect themselves (Bates et al., 2004; Crippen and Sheffield, 2006). Mainly considering that darkling beetles can maintain *Salmonella* internally during pupation, the next generation of beetles can recontaminate a poultry barn and the following flocks (Roche et al., 2009). Both fly and darkling beetle populations can explode within a few weeks, even with new litter (Fig. 2). Mites (*Dermanyssus gallinae*) have also been associated with recurring site contaminations with *Salmonella Enteritidis* (Moro et al., 2009) and *Salmonella Gallinarum* (Lee et al., 2020)

Insect control requires the application of biosecurity measures for the environment, equipment, visitors, and management of organic materials. It is called integrated pest management. First, the site should always be kept free of unnecessary material since these items can harbor vermin. From the outside, a stone access to the door of the building should be favored. Indoors, rodent traps, or bait should be set in the anteroom and regularly monitored. Insecticides should be applied to the ground, and equipment that

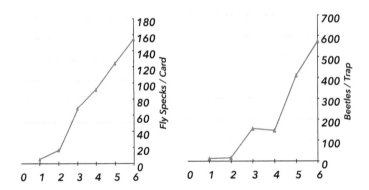

Figure 2 Growth of a fly and darkling beetle infestation in a turkey flock between placement and 6 weeks of age. The fly population was estimated using fly speck card (white card left 24 hours inside the barn and fly defecations counted); Beetle traps where located against the walls inside the barn; the number of beetles where counted after 24 hours.

must be in the anteroom should be off the floor as much as possible (Villa and Velasco, 1994). It is strongly suggested to rotate between insecticides to increase efficacy and coverage and avoid the development of resistance. Since the treatment is not selective, some recommend the application of insecticides only in areas with high larval density to avoid the destruction of predatory insects, which may be beneficial (Ebeling, 1975).

In addition, feeding and watering equipment should be regularly inspected and repaired to decrease insect populations. The feed delivery system and hoses should be periodically cleaned to prevent insect nesting. To reduce the risk of introducing certain insects, it is also necessary to control employee and visitor traffic and disinfect personal effects and any equipment entering a building. These measures are particularly important when the ectoparasites can survive outside the host for a few days to several weeks. Managing organic materials, such as bird carcasses and manure, is critical. Carcasses provide a substrate and a humidity level (over 90%) favorable for the development of fly larvae, and the manure allows infestation by beetles such as mealworms and their larvae (Allen and Newell, 2005; Barrington et al., 2006).

Rodents can also be found in large numbers on farms as long as water, food, and shelter are available. Rodents are mechanical and biological vectors of several pathogens. Indeed, the same serotypes of *Pasteurella multocida* are isolated in rats and poultry (Curtis et al., 1980). Like rats, mice are involved in the transmission of *Salmonella* (Henzler and Opitz, 1992; Davies and Wray, 1995). Mice are three times more infected with *Salmonella Enteritidis* than samples taken from the environment of a poultry farm. It seems that environmental contamination decreases in the absence of mice (Henzler and Opitz, 1992). Another study supports this view: there are three times more chances that the bacteria will persist in a building when rodents are present, despite washing and disinfecting the building (odds ratio (OR) 3.1; 95% confidence interval (CI) 1.1–3.2) (Rose et al., 2000).

Control of rats and mice is only effective if the barns are rodent-proof, that is, having no openings beyond 1.3 cm (Roy and Brown, 1954). It is also important not to attract rodents to farms by picking up any feed spills, not leaving debris on the site, and regularly pruning vegetation around buildings. Vegetation provides rodents with their basic needs (water, food), protection from predators, and material for building their nests. Appropriate pest management also includes traps or traps containing rodenticides and monitoring to ensure the effectiveness of the program (Axtell, 1999; Corrigan, 2006).

Other wild animals can be reservoirs and vectors of several diseases. Raccoons, along with opossums and foxes, appear to play a role in avian influenza outbreaks. During the 2002 epidemic in Virginia, a significant association was demonstrated between their presence on a farm site and the presence of the avian influenza virus in commercial chickens (OR 1.9; 95% CI 1.0–3.4) (McQuiston et al., 2005). On the other hand, it is more often migratory birds that are involved in the initial influenza outbreaks of an epidemic. Between 1978 and 2000, turkey farms in Minnesota experienced more than 100 introductions of low-pathogenic avian influenza virus from migrating ducks (Halvorson, 2002). Moreover, free-range poultry farms, more exposed to contact with wild birds, showed a higher risk of low-pathogenic avian influenza infection in the Netherlands between 2013 and 2017 (Bouwstra et al., 2017). Several highly pathogenic avian influenza H5N8 epidemics in poultry- and duck-producing regions in Europe between 2017 and 2021 originated from wild birds. Hence, there is clear evidence that wild birds serve as reservoirs, maintaining the virus in several regions over time (Jeong et al., 2014; Verhagen et al., 2021). The mobility of wild birds is a major concern in terms of disease dispersal. They frequently contaminate the environment, vehicles, equipment, feed storage, etc., with their droppings (Davison et al., 1997). It is, therefore, important to limit the access of wild birds and to avoid contact with domestic birds (Axtell, 1999). In that regard, free-range poultry productions require increased attention to biosecurity measures related to the outdoor range (e.g. type of vegetal cover, reducing the accessibility of feed and water to wild birds) (Bestman et al., 2018; Delpont et al., 2020).

It is best to avoid having pets on a farm. Virulent forms of *Pasteurella multocida* have been isolated from the oral cavity of farm cats (Van Sambeek et al., 1995). Strains affecting cats have also been shown to infect turkeys (Curtis and Ollerhead, 1982). There is little evidence of the role of cats in the direct transmission of the bacteria, except in cases of bites (Korbel et al., 1992). However, they can transmit the infection to rats and hunted wild birds, which can subsequently infect domestic poultry. For any type of production, it is therefore inappropriate to use cats for rodent control on a farm (Corrigan, 2006). Dogs, meanwhile, have been identified as a potential carrier of the avian infectious bursal disease virus (Pagès-Manté et al., 2004). The presence of other animals on the site of a broiler farm, including pigs, cattle, sheep, and poultry other than broilers, is strongly associated with a high risk of infection with *Campylobacter* (OR 6.33; 95% CI 1.54–26.00). Cattle are also an important source of introduction of this bacterium in broiler chickens via the farmers' boots (Van De Giessen et al., 1998).

6 Reducing sources of contamination: manure, litter and dead birds

6.1 Manure and litter management

Manure and used litter, and occasionally new litter, can be very contagious material, particularly of enteric pathogens, although it is well documented that a respiratory virus like the influenza virus can persist and spread easily via this route (Duvauchelle et al., 2013; Kim et al., 2018). Kim et al. (2018) reported a reduced risk of contamination when an outside company was contracted for manure removal. However, in Ontario, in 1995, growers who required an outside company to handle used litter were on average eight times more at risk of having a flock infected with infectious laryngotracheitis in comparison to neighbors located less than 2 km from them and having birds at risk at the same moment (Vaillancourt, 1995). Manure has also been identified as a source of contamination for *Campylobacter* (Arsenault et al., 2007).

Therefore, the handling of manure, storage, decontamination (composting, incineration), and removal conditions are of critical importance.

6.2 Disposal of dead birds

In many countries, rendering is the most prevalent approach to dead bird disposal. However, it has been associated with a greater risk of disease transmission in different species, including poultry. For example, in Virginia, United States, farms using rendering instead of disposing of carcasses on site were seven times more likely to have flocks infected with avian influenza H7N2 in 2002 (McQuiston et al., 2005). The proximity of the rendering container to poultry barns may partly explain the greater risk of contamination when renderers visit the site. Another important factor is the absence of an adequate sanitation protocol for personnel visiting the site where the rendering container is located as they return to the barns.

6.3 Incineration

Incineration refers to the combustion of material to the extent that the resulting end products are heat, gaseous emissions, and residual ash. There are three types of dead animal cremation: (1) stationary cremation, (2) air curtain cremation, and (3) open-air cremation. The overriding consideration affecting the use of on-farm incineration is regulatory. In some regions or countries, regulations state that incineration equipment must contain a secondary combustion chamber to reduce particulates (i.e. 'fly ash') and other emissions, such as polluting hydrocarbons and heavy metals (Chen et al., 2004).

Fixed plant units range from business units designed for on-farm animal cremation to large cremation plants. They are normally powered by diesel, natural gas, or propane. The equipment is relatively easy to operate after a short training. Periodic observation, routine maintenance, and ash cleaning are required. Fuel consumption varies depending on incinerator design and loading rate.

Air curtain incineration involves the use of air forced mechanically through a firebox with refractory panels or a trench in the earth. Each type has distinct characteristics that may increase or limit its potential for use as a carcass disposal method. Air curtain technology was developed primarily as a means of incinerating large amounts of combustible waste resulting from land clearing or natural disasters (Ellis, 2001).

The main feature of air curtain incineration is a high-speed 'curtain' of air produced by a fan above an above-ground combustion chamber or in an earth-burning trench. The air curtain serves to contain smoke and particles in the combustion zone and provides better airflow for warmer temperatures and more complete combustion. It was used for the disposal of dead birds during the avian influenza epidemic in Virginia (United States) in 2002 (Brglez, 2003). This is a fuel-intensive process (mainly wood and diesel fuel), but its use may be justified in order to avoid moving dead animals over long distances (Ellis, 2001).

Outdoor incineration is not recommended for several reasons. The downsides include labor and fuel requirements, reliance on favorable weather conditions, potential for environmental pollution, odors, and negative public perception.

6.4 Composting

Several studies indicate that composting appears to be effective in eliminating infectious pathogens endemic to poultry. In principle, large-scale windrow composting would be effective for the disposal of a large volume of carcasses.

There is little information on the fate of prions or sporulating bacteria such as *Bacillus anthracis* when composting cadavers, which prevents it from being considered an adequate method of managing mortalities by the European Union. Deactivation of prions is difficult: it requires exposure to heat from 980 to 1100°C or to alkaline digestion. Neither of these conditions occurs in the composting process. However, there is some evidence that certain enzymes and competition from organisms may have a beneficial effect in reducing the presence of prions during composting (Bonhotal et al., 2014). In contrast, in North America, unless there is a serious suspicion of prions, which is not the case for poultry, composting is considered an excellent means of eliminating the vast majority of infectious pathogens of interest to animal industries and government authorities. In other words, the advantages of composting, including the elimination of the movement of contaminated organic material, are considered far greater than the disadvantages.

Composting is a natural biological process of decomposing organic material in a predominantly aerobic environment. During the process, bacteria, fungi, and other microorganisms break down organic material into a stable mixture (compost) while consuming oxygen and releasing heat, water, carbon dioxide (CO_2), and other gases (Keener et al., 2000). The use of compost for the routine disposal of on-farm poultry carcasses has increased in prevalence in the United States and Canada over the past 30 years (Tablante and Malone, 2006).

Four variables are considered essential for successful composting: (1) moisture content (40-60%), (2) temperature (45-60°C); (3) oxygen concentration (desirable level 10%), and (4) carbon:nitrogen (C:N) ratio (desirable range 20:1 to 30:1) (Keener et al., 2000).

The process essentially takes place in two phases - a primary thermophilic phase (temperatures up to 70°C) and a secondary mesophilic phase (usually 30-40°C) (Kalbasi et al., 2005).

Temperature and temperature maintenance are important factors in the destruction of infectious pathogens. A temperature of 54°C for 3 days, typical of carcass composting, should kill all pathogens except spores and prions (Sander et al., 2002).

To keep costs down on the farm, composting of carcasses is usually done by producing a static pile or heap that does not involve specialized mixing, crushing, turning, aeration, and screening equipment. The degradation of the carcass is initiated by natural anaerobic bacteria in the cadaver and by aerobic bacteria on the exterior surfaces. Odorous gases and liquids diffuse into drier, more aerobic plant materials, where they are ingested by microorganisms and broken down into simpler organic compounds and ultimately into CO_2 and water (Keener et al., 2000). The success of this operation relies on the careful construction of a layered heap using appropriate amounts of plant-based covering material below, between, and above the carcasses. Characteristics of effective toping materials include water-holding capacity, gas permeability or porosity (oxygen for microbial activity), biodegradability, wet strength, and sufficient carbon. These physical characteristics determine the ability of toping materials to absorb excess liquids, preventing the release of leachate and odors (King et al., 2005).

A variety of plant-based residues have been used as toping materials, including sawdust, wood chips, ground corn stalks, straw or ground hay, oat or peanut hulls, poultry or livestock bedding, dry manure, etc. (Keener et al., 2000; Glanville et al., 2006).

Turning a stack may be necessary to break up wet areas and to introduce more oxygen and moisture, if needed, to reactivate aerobic microbial activity and stimulate a secondary cycle of heat production. Once the secondary heating cycle is complete, soft tissue decomposition is usually complete and the compost is stable enough to be stored prior to field application. Based on a review of the literature, Keener et al. (2000) concluded that decomposition times are largely a function of cadaver mass, and they published weight-based prediction equations for the duration of the primary and secondary composting cycles. In practice, most of the compost is turned only once or twice. Turning accelerates the decomposition of the carcasses, but it is not essential if the carbon source used to cover the carcasses is sufficiently permeable for the diffusion of oxygen into the heap (Glanville et al., 2006).

On-farm composting is usually done in open bins or swath piles. Three-sided bins are typical, with the open end allowing access for placement, turning, and removal of compost using a tractor. Permanent structures are built with treated lumber or concrete and are usually built on a concrete platform to provide a firm work surface. Windrow composting involves the construction of long, narrow piles having a parabolic or trapezoidal cross section. Due to their shape, windrows have a large exposed surface that encourages aeration and drying (Mukhtar et al., 2004). Since the dimensions of the windrows are not constrained by the walls, their dimensions can be adapted to any size and number of carcasses, which makes them particularly useful following the mass slaughter of animals.

Although windrows do not require the construction of a structure to contain the compost, a low permeability base is recommended to avoid contamination by leachate from the underlying soil (e.g. concrete or asphalt; gravel lined with fabric plastic or geotextile; compacted soil) (Keener et al., 2000).

7 Separating healthy birds from sources of contamination: zoning production sites

Basically, zoning is about preventing the contact between susceptible birds and microbes. The barns where the birds are located must be the most restricted zone (restricted access zone or RAZ); the farm where we find the barns is a controlled access zone (CAZ) (Fig. 3). Everything else around these two zones must be considered an environment that is source of contamination (Anon., 2018).

7.1 Preserving the integrity of the controlled access zone

All farm personnel must avoid live bird markets and be in contact with any other poultry. The CAZ should be clearly marked or understood by those who must access the site. A fence is best when economically doable. The main point is that only essential vehicles, personnel, and equipment should get within the CAZ.

An integrated pest management program, including cleaning out vegetation for at least 5 m around the barns, is paramount to removing potential carriers of disease (Vaillancourt and Martinez, 2001). Standing water must be avoided on the premises. It is a breeding environment for insects, and it attracts wild birds that may be carriers of infectious pathogens such as avian influenza. Feed spills should also be avoided for the same reason. If rendering is used to dispose of dead birds, the disposal area cannot be part of the CAZ (Anon., 2018).

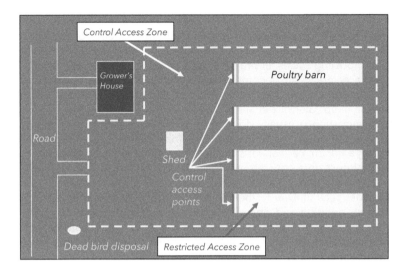

Figure 3 Control access zone (area where poultry traffic and equipment are controlled; delimitation can be virtual or physical (e.g. fence); and restricted access zone (area where birds are located). The control access points are the doors, and ideally the anteroom, providing access to the flock.

Finally, equipment and vehicles must be washed and disinfected, most importantly, all equipment used on more than one farm. In this case, it is best to clean and disinfect when leaving the farm and prior to use on the next farm (Anon., 2018).

7.2 Protecting the restricted access zone

The purpose is to limit access to essential visitors. When a visit is needed, it is important to make sure that visitors have taken a shower and changed clothing and footwear before visiting. Even if they have not been on another farm before, they may still have had an opportunity to get contaminated on their way to the production site. Employees and even the farm owner should consider themselves visitors when they leave the facilities and return. It is best for them to wear farm-specific clothing (laundered each day or supplied at the farm) and boots (Anon., 2018).

7.3 Anteroom: managing barn entrances

The separation between the RAZ and outside must be clearly delineated. When a shower is present, the shower itself is the line separating the two. In the absence of a shower, it is important to delineate the zones with at least a line. Anterooms separated into only two zones (external or contaminated and internal or clean) can be problematic because cross-contamination between the two zones is frequent (Figs 4 and 5). Therefore, keeping the barn entrance free of any organic matter will prevent cross-contamination of these two areas. However, it is difficult to achieve without a physical barrier between the zones. In

Figure 4 Equally dirty contaminated (outside) and clean (inside) zones separated by a red line in an anteroom.

Figure 5 Examples of two-zone barn entrances. Note that a hydroalcoholic solution dispenser for hand decontamination is located at the level of the separation between the zones. The red line design is less effective for compliance than the ones with a bench.

addition, when designing an anteroom, it is recommended to have separate drainage for each area. Finally, an easier and more effective anteroom design is one with three zones (Fig. 6). Often known as the Danish entrance, the three-zone design offers an area between

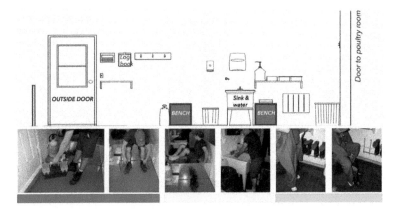

Figure 6 Layout of a three-zone barn entrance and the steps that must be followed to pass from the red zone (contaminated area) to the green zone (clean area) via the yellow zone (intermediary area for hand washing).

the contaminated and clean zones where visitors and personnel can wash their hands. This step, while transitioning between the contaminated and clean zones, is important because it ensures that hands are decontaminated before handling boots, clothing, or equipment on the clean side of the anteroom (Anon., 2018).

In an anteroom, porous surfaces (concrete, wood) are generally more prevalent than smooth surfaces (plastic, metal). For detergent application on porous surfaces, it is recommended to use a low-foaming detergent. For disinfection, the use of fumigation (0.5 μm particle) might be preferable for its ability to reach corners and penetrate cracks (Battersby et al., 2017), although research is inconclusive on this point.

The type of soil also has an impact on the sanitation of an anteroom. For example, a cracked concrete floor, a porous soil where water stagnates easily, will be much slower and more difficult to disinfect compared to a floor covered with flat epoxy. Stagnant water is a source of multiplication for microorganisms and should be given special attention.

7.4 Footbaths

The effectiveness of a footbath is influenced by several factors. The two major factors are the presence of organic matter (on the boots or in the footbath; Fig. 7) (Amass, 1999; Amass et al., 2000; Rabie et al., 2015) and the often insufficient contact time. Other factors also influence the effectiveness of footbaths: the type and concentration of the disinfectant (Amass et al., 2000; Hauck et al., 2017), the location (Fig. 8), the pH and the hardness of the water, and the material of the container used (e.g. metal container). For example, if a footbath is left outside, the disinfectant may be inactivated by UV rays from the sun (e.g. iodine), diluted by rain, and affected by warm (evaporation) and cold (freezing) temperatures (Rabie et al., 2015).

Despite all these issues, some epidemiological studies have shown a protective effect of footbaths, especially in connection with contamination with *Campylobacter* spp. and with avian influenza H9N2 (Chaudhry et al., 2017). However, these results are not

Figure 7 Examples of footbaths left outside and rendered ineffective due to a large amount of organic material.

supported by other studies (Refrégier-Petton et al., 2001). Allen and Newell (2005) report a lower efficiency of footbaths compared to using boots dedicated to each barn.

Beyond epidemiological studies, other studies have evaluated the direct effectiveness of footbaths against infectious pathogens under different conditions (contact time, presence of organic matter, wet or dry prewash, type of disinfectant, etc.). In studies demonstrating good efficacy, Dee et al. (2004) reported that a sodium hypochlorite footbath (bleach) allowed the complete elimination of the PRRS virus with a contact time of 5 s in the disinfectant solution. It is important to note that the footbath was replaced after each use (single use), which is not done in practice. Amass et al. (2006) reported a slight reduction, insufficient to be considered significant, in bacterial contamination on lightly soiled boots when passed over a disinfectant mat. Morley et al. (2005) reported a 67–78% reduction in bacterial load after 7 min of contact time with the disinfectant solution on lightly soiled boots. Finally, Dunowska et al. (2006) reported a reduction in bacterial load of 95.4–99.8% after 10 min of contact on lightly soiled boots. The studies reporting certain effectiveness can thus be criticized because the reduction of the microbial load is inadequate, the solution needs to be replaced after each use, or the contact time required is unrealistic in practice.

Several studies have confirmed the inadequacy of traditional footbaths as a biosecurity measure. Hauck et al. (2017) reported that quaternary ammonium and quaternary ammonium-glutaraldehyde footbaths were not able to eliminate highly pathogenic (H5N8) and low-pathogenic (H6N2) influenza. Several studies have demonstrated that the presence of organic matter prevents the effective disinfection of boots, regardless of the contact time (Amass et al., 2000; Rabie et al., 2015). In fact, the mechanical action of a brush to remove organic matter from the boots is even more effective than walking through a footbath without prewashing. It is important to note that a footbath can even be a risk factor (Amass et al., 2000) because it provides a humid environment favoring bacterial growth and sharing of pathogens between treated boots. Having this in mind,

Figure 8 Examples of footbaths positioned in such a way as to prevent users to march forward as they should between two zones.

Owen and Lawlor (2012) compared phenol- and quaternary ammonium-filled footbaths to dry footbaths filled with a dry bleach powder. The average residual life of the liquid footbaths was less than 2 h and did not significantly impact the bacterial count on the boots. By contrast, the dry bleach footbath reduced the bacterial count by around 98% and had a residual life of 14 days.

7.5 Downtime

Downtime is the period when a poultry barn or an entire production site is without a flock. The concept is to remove birds in order to apply sanitation measures. Repopulation follows a few days to a few weeks later. This is called all-in/all-out (Anon., 2018). This approach has a protective effect on *Salmonella* contamination (Snow et al., 2010) and avian influenza (Kim et al., 2018). In a study on coliform cellulitis in chickens in Canada, the longer the downtime, the lower was the prevalence of cellulitis-related condemnations at slaughter (Elfadil et al., 1996). Chin et al. (2009) reported that an extended downtime of 30–91 days contributed to the regional control and eradication of infectious laryngotracheitis in California. Field experts participating in a Delphi study indicated that a 14-day downtime between meat bird flocks was the norm, with 21 days between breeder flocks (Vaillancourt and Martinez, 2001).

8 Separating healthy birds from sources of contamination: hatchery

There is more to hatchery biosecurity than just sanitation. Hatchery location, design, accessibility, workflow, and pest control all complement a stringent sanitation program to ensure good biosecurity. The temperatures and humidity required for hatching are ideal for the growth of bacteria and molds. Problems coming from the breeders can be amplified, and new ones introduced (Kim and Kim, 2010).

Embryos and newly hatched chicks can offer little disease resistance. In other words, the pressure of infection required for affecting their health and viability is much lower than for older birds. It is, therefore, essential to provide them with a clean environment during their incubation and hatching (Bennett, 2017).

The more a hatchery is located in close proximity to other poultry traffic, the more vigilant one should be about issues such as pest infestation, visitors, and ventilation-related pathogens. It is important to recognize that a lot of poultry in the area increases the risk of transmission of infectious diseases. Therefore, it is best if no poultry-related activities (feed mill, farms, live-haul routes, etc.) are conducted next to the hatchery (Bennett, 2017).

8.1 Personnel

All employees must park in a designated area. They should not be visiting other poultry facilities. On the rare occasion when this might occur, they would have to take a shower and use clean clothes before reentering the hatchery. They should use protective clothing, including boots, only used at the hatchery. If a shower in-shower out facility is available at the hatchery, it must be used consistently and never by-passed. At the very least, all employees will be expected to have taken a shower at home before coming to work, they must come to work wearing clean clothes, and they will all be required to thoroughly wash their hands when arriving at the hatchery. Therefore, a handwashing station with antibacterial soap should be readily available. Handwashing is essential before, after, and between egg and chick or poultry handling operations (e.g. setting, transfer, candling, sexing, vaccinating, packing, etc.) (Thermote, 2006; Bennett, 2017).

8.2 Hatchery workflow

The work and traffic flow should follow the same route as the hatching egg. Modern hatcheries should be designed with ventilation systems to prevent cross-contamination of the different areas of the building. The benefits of such systems will largely be negated if employees are allowed to move freely back and forth between separate areas of the building. If it is needed to 'break' the unidirectional workflow, it is important for all employees to wear clean hatchery clothing and to change it as necessary. The hatchery workflow is as follows: egg receiving area, egg holding area, egg cooler, setters, hatcher rooms, tray dumping, chick processing area, chick holding area, and chick loading area (Mauldin, 2002).

The key is to minimize contamination from one room to the next. Positive pressure rooms are important in critical areas so that contamination will not be drawn in through an open door. Doors help stop cross-contamination between rooms (Mauldin and MacKinnon, 2009).

8.3 Egg delivery

The driver cannot enter the egg holding room at the hatchery, in particular, if several breeder farms are visited. After delivery, the egg truck is washed and disinfected. Clean and disinfected trays and trolleys are returned from the hatchery to the breeder farms (Bennett, 2017).

8.4 Egg sanitation

Eggs are not laid in a sterile environment. Even before that, eggs are exposed to many microbes. The vent of the hen is a contaminated 'delivery environment'. Dirty nests, or if eggs are laid on the floor, can also contribute to egg contamination. Eggshell surfaces must be kept dry. Without humidity, most microbes will have a hard time multiplying or penetrating the egg's natural defenses. Therefore, it is important to avoid the sweating of eggs, such as by moving them from a warm to a cool environment. Bacteria inside an egg will use some of its nutrients and may affect the growth or even kill the embryo. Contaminated eggs in incubators and hatchers can also serve as a source of contamination for the other eggs and newly hatched birds.

Early embryos can also be affected by chemical vapors. So the choice of sanitizing products in the egg reception and storage areas is important (please see the following text). Eggs that have been disinfected on farm can be moved immediately to setter trays for storage. Eggs of unknown sanitation status normally should not be allowed in the hatchery. If the decision is made to allow them, they must be disinfected before setting. If dirty eggs are found (obvious presence of fecal material, litter, and feathers), they should be removed and held for breeder management to review.

Egg sanitizing (Ernst, 2004): It must be done properly; otherwise, the result may be a greater degree of egg contamination. It is important to keep sanitizing water hotter than the eggs (43-49°C). The sanitizing solution must contain a detergent sanitizer. It is best to use a washer that does not recirculate water.

Chlorine and quats (quaternary ammonium) have been excellent disinfectants. Quats offer residual activity, are safe for the operator, are relatively cheap, and are compatible with antibiotic dipping. Quats are safe for hatching eggs up to 10 000 ppm but are effective as a disinfectant at 250 ppm (large safety margin) with 10 ppm ethylenediaminetetraacetic acid. It is best to keep the solution at a pH of 8 for optimum efficacy. This can be achieved by using sodium carbonate (Rodgers et al., 2001).

Hatching eggs can also be sanitized by fumigation using 1.2 mL of formalin and 60 g of permanganate sulfate per cubic meter in a disinfection cabinet. Other products may be used for fumigation, and new technologies are also considered, such as pulsed ultraviolet light (Cassar et al., 2020) and low-energy electron beam (Steiner, 2020).

8.5 Building and equipment sanitation

Washing and disinfection of equipment are a critical component of hatchery biosecurity. A thorough cleaning of the area (setters, hatchers, floors, chick-go-rounds, vaccinators, etc.) is essential before a disinfectant can be applied. Organic material (fluff, blood, shells, and droppings) reduces the effectiveness of disinfectants. Therefore, washing is paramount.

8.6 Setters

Fog a disinfectant after eggs have been set. Multistage setters should be fogged each time new eggs are set or transferred. When setters are emptied, a thorough cleaning and disinfection are required. Like for any other areas, all debris must be removed as part of the cleaning process.

8.7 Hatchers

The hatcher is the main source of organic contamination in the hatchery (eggshells, unhatched eggs, dead chicks/poults, fluff, droppings). It is, therefore, important to use equipment handling and sanitation procedures to minimize contaminating the other areas of the hatchery.

8.8 Bird processing equipment

This equipment requires a very rigid sanitation schedule to avoid bird quality problems (Mauldin, 1983). Special attention should be paid to the vaccination equipment (syringes, vaccine containers, and hoses washed with a nonresidual disinfectant such as alcohol or chlorine).

Chick-Go-Rounds and/or chick/poultry conveyor belts and other processing equipment can contaminate day-old birds. At greater risk are chicks or poults that may have been pulled early. If their navel is incompletely closed and comes in contact with a dirty or contaminated conveyor belt, there will be many cases of infection. Therefore, it is important to clean and disinfect conveyor belts after the chicks/poults from each breeder flock are processed.

8.9 Ventilation

The hatchery ventilation system plays an important role in preventing contamination (when functional and well designed) or by being the source of the problem (e.g. aspergillosis contamination). Ideally, each room of the hatchery should have its own separate ventilation system so air does not move from one room to another. This way, the ventilation system will not introduce contaminated air from other sections of the hatchery or from outside. The system must provide clean, fresh air to the hatchery and the incubators at all times. Hatcheries should be designed to exhaust air away from intake outlets, so contaminated air is not recycled into the hatchery (Mauldin, 1983).

Maintenance of the evaporative coolers is critical. Filters must be cleaned and disinfected routinely. Disinfectants can be added to the water in the evaporative coolers, especially when starting them off in the spring.

8.10 Monitoring

Records must be kept that the biosecurity program is in place and implemented correctly and continuously. The hatchery should have a quality control program to monitor incoming eggs visually and microbiologically. A continuous monitoring program of the building and equipment must be in place to determine the microbial populations in the hatchery (Racicot et al., 2020). Samples may be collected from the tray wash area, air intakes and outlets, filters, evaporative coolers, setters, hatchers, air in chick holding and egg storage rooms, chick conveyor belts, water source to hatchery, vaccination equipment, including vaccine and diluent (Mauldin, 1983). Samples must be tested for bacteria (coliforms, *Salmonella*) and fungus.

9 Separating healthy birds from sources of contamination: regional biosecurity

The relationship between proximity to a poultry production site and the probability of contamination has been well established. When the distance between two egg layer farms increases by 1 km, the risk of colibacillosis is six times less in Belgium (Vandekerchove et al., 2004). Egg layer farms are also twice less likely to be contaminated with *Salmonella* if the nearest farm is at least 1 km away (Snow et al., 2010). The risk of infectious bursal disease (Gumboro) transmission was higher when the susceptible flock was less than 20 km away from an infected flock. Poultry traffic, wild birds, and airborne transmission might have contributed to such disease transmission in Denmark (Sanchez et al., 2005). In Australia, East et al. (2006) demonstrated that the risk of Newcastle disease infection is reduced threefold in egg layer farms if the nearest farm is at least 10 km away; fourfold for breeder farms if the nearest poultry farm is at least 1 km away, and threefold in broiler chickens if at least 500 m away. A poultry farm located less than 1 km away from an avian influenza-infected flock is about 35 times more at risk of infection compared to a farm located at least 10 km away (Boender et al., 2007). Commercial poultry farms infected with the infectious laryngotracheitis virus were 36 times more likely to be within 1.6 km from a backyard flock than disease-free farms (Johnson et al., 2005).

In Australia, veterinary authorities, in collaboration with the poultry industry, have established guidelines for the geographic distribution of poultry farms (Table 1). Although these are not yet mandated by law, they were determined and accepted collectively based on scientific evidence. Note that these are standards for regular poultry operations.

Table 1 A guide on biosecurity buffer distances[a]

Farm type	Species	Buffer (m)
New farm	Fowls/turkeys/other avian species, e.g. ratites, quail	1000
Units in large farm complex	Fowl/turkeys or other avian species	200-500
Farm complexes	Fowl/turkeys or other avian species	>2000
Breeder farms	Fowl/turkeys or other avian species	2000-5000
Duck or waterfowl farms	Duck, waterfowl	5000

[a]The buffer is measured from either the nearest barn walls for older type building or from the centroid of the mechanical ventilation system of the newer tunnel ventilation barns. Reference: http://new.dpi.vic.gov.au/notes/agg/poultry--and--other-birds/ag1155-biosecurity-guidelines-for-poultry-producers

The vast majority of infectious diseases are horizontally transmitted. Cross-traffic between farms and the joint use of equipment are certainly significant modes of disease transmission (Guinat et al., 2020). With potential vectors such as flies and darkling beetles, known vectors of avian influenza, disease control efforts in regions of high farm concentration cannot rely solely on farm-level measures. In a competitive environment, it is tempting to limit data sharing. Although liability will always be a concern, pointing fingers has never been an effective disease control strategy, and poultry organizations sharing a region must also share the necessary information needed to prevent and contain significant contagious diseases. A regional approach to biosecurity includes establishing separate geographical working zones for service personnel. When an important disease is suspected on a farm, a self-imposed quarantine matched with communication with key personnel will go a long way to prevent a major epidemic (Vaillancourt et al., 2018).

Communication remains central to the success of regional biosecurity, especially when an outbreak occurs. In 2005, during an outbreak of infectious laryngotracheitis in California, two integrators coordinated to perform an extended downtime in a region and to set up biosecurity audits. Cooperation and communication between companies have helped eradicate the disease. The importance of communication is also highlighted in a 2001 survey of 72 North American veterinarians on the importance given to certain elements of biosecurity (Vaillancourt and Martinez, 2001).

10 Biosecurity compliance

Biosecurity compliance is defined as the behavior of a person coinciding with the recommendations of professionals in the control and prevention of infectious diseases. The measure of compliance is reported as the ratio of the number of recommendations applied to the number of recommendations prescribed; in other words, the ratio of the number of biosecurity measures applied to the number required. In practice, this is expressed as a percentage.

The constant application of biosecurity measures is essential for the success of any type of animal production. However, compliance with biosecurity measures is sporadic and variable regardless of the type of production, including pig and poultry productions (Losinger et al., 1998; Pinto and Urcelay, 2003; Boklund et al., 2004; Ribbens et al., 2008).

Similar observations are made in human medicine. Indeed, although compliance with hygienic measures among physicians is relatively high (80%) during high-risk procedures, this is not the case (30%) for low-risk procedures. In addition, it is interesting to note that compliance with medical recommendations by patients (33–54%; Haynes et al., 1979) also corresponds to what is observed in commercial poultry farms (Racicot et al., 2011).

There is a weak correlation between reported compliance and observed compliance ($r = 0.21$) for handwashing (O'Boyle et al., 2001). In veterinary medicine, few studies report compliance as observed. When it does, the results are worrying. For example, a procedure to register visitors to turkey farms in North Carolina was filmed on three farms shortly after the procedure was put in place. Compliance ranged from 7% to 49% (Vaillancourt and Carver, 1998).

A study in Quebec on compliance with biosecurity measures in the anteroom of poultry production buildings on eight farms was carried out using hidden cameras (although stakeholders on the farm were informed that they would be filmed). A total of 44 different errors were observed from 883 visits made by 102 different people. On average, four errors were recorded per visit. The maximum number of errors made by a person during a visit was 14. Twenty-seven of the 44 errors (61.4%) were related to measures relating to the respect of the zones (clean or internal versus contaminated or external), 6 for boots (13.6%), 5 for handwashing (11.4%), 3 for coveralls (6.8%), and 3 for visitation records (6.8%). The nature and frequency of errors suggest a lack of understanding of biosecurity principles (Racicot et al., 2011).

Since compliance with biosecurity is generally weak, it is essential to establish strategies to improve the implementation of biosecurity measures. A study evaluated the value of audits and visible cameras on compliance with the biosecurity measures required when entering and leaving poultry barns of 24 farms in Quebec. The evaluation was done initially during the first 2 weeks after the intervention, as well as 6 months later for another evaluation lasting 2 weeks. Nearly 2800 visits by 259 different people were recorded on video. The results showed that bimonthly audits had no impact on compliance in the medium term, 6 months after the start of the project. It should be noted, however, that the auditor was not in a position of authority. The visible cameras had an impact on the change of boots (OR = 9.6; 95% CI 1.9–48.4) and respect for areas (contaminated versus clean) during the visit (OR = 14.5; 95% CI: 1.2–175.1) for the short-term period. However, 6 months later, compliance declined and was no longer significantly different from control farms. The duration (>5 min) and the time of the visit (morning), the presence of the producer or an observer (sometimes negative effect), the design of the anteroom (in relation to the ease of application of the measures), the number of buildings (more than five buildings), the number of biosecurity measures requested, the type of boots worn (plastic boots), and being a member of the grower's family (negative effect) were significantly associated with biosecurity compliance. The same study also showed that certain personality traits were associated with compliance, as well as the number of years of experience in poultry production and the level of education of the participants. For the number of years of experience, the relationship was not linear. It is possible that early-career circumstances (e.g. avian influenza epidemic) could have a long-term impact on compliance (Racicot et al., 2012b).

Delpont et al. (2020) attempted to determine how sets of socio-psychological factors (i.e. knowledge about biosecurity and transmission of avian influenza, attitudes, personality traits, social background) affect the adoption of on-farm biosecurity practices. The study was carried out as part of 127 visits to duck farms in southwestern France. The factorial analysis of mixed data and the analysis of hierarchical grouping identified three groups of producers with different socio-psychological profiles. The first group was characterized by minimal knowledge, negative attitudes toward biosecurity, low social pressure (defined as a concern to preserve the avian influenza free status of the farm in order not to tarnish one's reputation), and a low level of conscientiousness (conscientiousness is a fundamental personality trait that reflects the tendency to be responsible, organized, hardworking, goal-oriented, and adheres to standards and rules). The second group was characterized by greater experience in poultry production, increased stress (excessive nervous tension), and social pressure. The third group was characterized by less experience in poultry production but better knowledge and positive attitudes toward biosecurity, increased self-confidence, and a focus on action. The first group had significantly lower adoption of biosecurity measures than the other two groups.

The measures implemented on farms vary greatly, depending on the animal species, type of production, region, stocking density, and possibly regional health status. For example, a Canadian survey (Young et al., 2010) of 642 broiler breeder owners indicates that 13% of respondents do not allow access to visitors. Of the rest, 36% require them to wash their hands and 50% require additional protective clothing to be worn before entering a barn. The main reasons for not requiring visitors to wash their hands were the lack of perceived need (20%), the time required by this measure (20%), and the lack of facilities (18%). Almost a third of farm owners also did not require the use of coveralls because of the time required to put them on, and 20% found it sufficient to wear only boots.

The known risk of disease transmission could favor the better implementation of measures (Dorea et al., 2010). The implementation of biosecurity measures also depends on the individuals being questioned. Indeed, producers and technical staff responsible for health monitoring do not always agree on what should be done and even what is done on a given farm. For example, in a study carried out in Canada, a weak to slight agreement was observed when these technicians were asked about the restrictions required by the farm owner to have access to the farm and their responses were compared to those of these owners (Nespeca et al., 1997). It is therefore important to provide each employee with written plans for the required biosecurity measures and to ensure a continuous training program (England, 2002).

10.1 Obstacles to compliance

Several reasons are given to explain the lack of compliance with biosecurity measures. Lack of knowledge or understanding of measurements is the main reason (Lotz, 1997; Barcelo and Marco, 1998; Sanderson et al., 2000). Lack of communication, time, incentives (positive and negative) to follow the rules, lack of audit programs, apathy or denial of potential risks, and economic constraints are also considered to be important factors (Vaillancourt and Carver, 1998). Communication is particularly necessary during epidemics. In 2005, during an outbreak of infectious laryngotracheitis in chickens in California, USA, two integrated companies coordinated to carry out an extended downtime and implement extensive audits. Cooperation and communication between companies have helped eradicate the virus (Chin et al., 2009). Gunn et al. (2008) also underline the importance of better collaboration between producers, veterinarians, and technical staff working on farms. To ensure good communication, the messenger is essential, but the content of the message is decisive. Several private and public organizations produce training materials for producers, but their content varies widely. This lack of harmony between training programs, and the resulting confusion, likely contribute to the lack of application of biosecurity (Jardine and Hrudey, 1997; Moore et al., 2008).

For a given disease, the perceived risk of infection influences the implementation of biosecurity measures. Unfortunately, the perception of risk differs greatly between producers, making consistency in the application of biosecurity measures difficult. The perception of risk is strongly linked to the understanding of the principles of biosecurity and the regional incidence of diseases. Perception of risk is also influenced by the intensity of traffic on the farm. Thus, producers of small farms or family farms often consider themselves less at risk compared to producers of intensive commercial livestock farms (Larsen, 2009).

In addition to the perception of risk, other individual factors may influence the application of recommendations. These are the attitudes, personality traits of the individual, experience with the disease, level of education and personal beliefs about animal health, incidence, prevention, and disease control (Buckalew and Sallis, 1986; Strömberg et al., 1999; Delabbio et al., 2005).

10.2 Intervention strategies

Haynes et al. (1979) describe different strategies to improve adherence to medical recommendations. Reminders of recommendations increase compliance from 24% to 70%. Thus, proper instructions with clear information and repeated feedback to stakeholders should improve compliance.

When it comes to handwashing, studies show that increasing accessibility to washing stations, training programs, and frequent feedback significantly increase compliance. This can go from less than 20% to more than 70% after the establishment of a training program and a monitoring system with frequent feedback (Tibballs, 1996; Colombo et al., 2002). However, compliance declines over time. This deterioration appears to be multifactorial and probably associated with the lack of a continuing education program (Conly et al., 1989).

10.3 Motivation of farm personnel

Motivation is another important element. Compliance with industrial safety rules is intimately linked to the motivation of employees to comply (Vroom, 1994). Clear goal statements and performance feedback have an optimal motivational effect when these two elements are combined. To increase motivation toward desired behavior such as compliance with biosecurity measures, the ability to achieve goals and a sense of self-efficacy are important (Bandura and Cervone, 1983). Self-efficacy is defined as 'the ability to implement the behavior essential to obtain a given result' (Pervin and John, 2001). According to Bandura and Cervone (1983), 'Our ability to deal with a situation and control the outcome is the key element that actually influences behavior. It is by manipulating the feeling of self-efficacy (e.g. by means of feedback announcing to the person that, compared to the performance obtained by others, their own performance is very good), rather than by giving feedback regarding a risky behavior, that a change in behavior is likely (Pervin and John, 2001).

Millman et al. (2017) have clearly demonstrated the perverse effect of a contradiction between what is required and what can be achieved. They studied the behavior of poultry catchers who had to follow several biosecurity procedures when the time allowed to do so was not sufficient. This situation led catchers to openly question the protocols and the need to respect biosecurity measures. Their noncompliance with the requirements was then seen as resourcefulness in the face of unrealistic conditions. Timmermans and Berg (1997) also stress the importance of 'local universality'; that is, standardized practices, such as biosecurity measures, can only be universal (achievable in different places and times) if they can be adapted locally.

10.4 Training program

A training program should include an evaluation of the participants before the training (theoretical and practical parts) and at the end. It is necessary that the trainer be credible and have the confidence of the participants so that the message is delivered well. Participants should also leave the session with a copy of the training material. A website accessible to participants will include materials, as well as program updates, upcoming training schedules, and links to other sites providing information on biosecurity. Conly et al. (1989) showed that training programs are not sufficient to maintain long-term compliance. Feedback and reinforcements (incentives) are needed. Reminders in various forms are also useful, such as messages in agricultural journals, mailings, contests and awards at annual meetings (Bradley, 2007).

10.5 Audits

Three audits over a 6-month period had no impact on biosecurity compliance in eight poultry farms in Canada (Racicot et al., 2012a). So it is not always necessarily impactful. In order to be valid, the audit process must be relevant, objective, quantifiable, repeatable, and capable of suggesting changes to be made. Standards must be clearly identified. These standards must reconcile current practices with protocols already defined, evaluated, and published. The data must collectively be the subject of regular reports showing the points assessed, the improvements or shortcomings identified, the corrective measures and their results. Finally, audits must be subject to independent evaluation to improve the process (Shaw and Costain, 1989; Smith, 1990). When these conditions are met, on-farm biosecurity audits have been shown to be very effective in aquaculture in New Zealand (Georgiades et al., 2016). In Belgium, an audit system, Biocheck.UGent, has been developed for swine and poultry and has made it possible to establish a link between the level of biosecurity and production performances (Gelaude et al., 2014; Rodrigues da Costa et al., 2019).

11 The economics of biosecurity

There is a paucity of scientific articles directly investigating the economic impact of biosecurity. An economic assessment of biosecurity in broiler breeders in 1987 indicated that a benefit–cost ratio of at least 3 should be expected for a farm considered at a 30% risk of being infected by an agent causing a severe disease (Gifford et al., 1987). Morris (1995) states that the primary purpose of any business should be to maximize return on investment over the long term. This is an important concept because a comprehensive biosecurity program does not necessarily offer a quick pay back. Yet, it is, or should be, an essential component of a farm owner's long-term strategy for success.

Biosecurity investments involve a mix of fixed and variable costs. Knowledge of costs would help inform cost-sharing programs related to animal disease mitigation efforts (Pudenz et al., 2019). In British Columbia, shortly after the H7N3 outbreak in 2004, these two types of costs were reported to represent about 4.5% of the cost of production for duck farms (JP Vaillancourt, pers. Comm.). However, in determining the economic impact of biosecurity measures, one must also consider the benefits in terms of reduced losses associated with infectious diseases, including the reduction in the use of antibiotics (Rojo-Gimeno et al., 2016). In pig production, a negative correlation between enzootic pneumonia, pleurisy, acute pleuropneumonia, and internal biosecurity underscores the importance of good biosecurity in reducing health problems (Pandolfi et al., 2018). In South Africa, financial modeling over a 3-year period of a farrow to finish pig farm of 122 sows estimated that, in the absence of disease, the implementation of biosecurity resulted in a 9.70% reduction in the total annual profit. In contrast, the study found that the implementation of biosecurity and its effective monitoring would prevent losses due to Africa Swine Fever with an impressive benefit–cost ratio of 29 (Fasina et al., 2012).

The perceptions of pastoralists on the cost of biosecurity are important for the adoption of these measures. If the perceived costs are excessive, producers may prefer not to apply the measures. This is in accordance with Casal et al. (2007), who found that perceptions about biosecurity measures and their use were interrelated. Other studies such as Fraser et al. (2010), Valeeva et al. (2011), and Toma et al. (2013) also found that the costs and benefits of biosecurity contribute to the adoption of biosecurity measures (Niemi et al., 2016). But the likelihood of adopting biosecurity increases inelastically as perceived costs decrease. This means that the relative change (percent) in biosecurity adoption is smaller than the relative change in assumed cost. On the other hand, if the grower receives information about a low-cost biosecurity technology that is available, it increases the rate of biosecurity adoption.

12 Conclusion and future trends

We essentially know the risk factors associated with infectious disease transmission in poultry. Biosecurity measures have been developed over the years. Two main challenges require research for on-farm measures: (a) finding ways to make specific measures easier, faster, and cheaper to perform; (b) how to increase compliance. Any progress in point (a) will favor point (b). But this second point has been associated with many factors, namely, lack of knowledge; economic constraints; lack of training, communication, incentives, time; difficulty in applying the requested measures; lack of audits; lack of consistency in the information available; beliefs, attitudes, perceptions, education, experience, and personality traits of farm workers. It is essential to consider several of these factors at the same time if one wants to have a significant impact on biosecurity compliance.

Finally, research is needed regarding regional biosecurity measures, such as the management of poultry traffic, zone raising (regional downtime), and zoning.

13 Where to look for further information

Good introductions to the subject for non-specialists:

- Dewulf, J. and Van Immerseel, F. (Eds). (2019). Biosecurity in Animal Production and Veterinary Medicine. CABI. 523 pp.
- Owen, R. L. (2011). *A Practical Guide for Managing Risk in Poultry Production* (No. V480 OWEp). 276 pp.
- Racicot, M., Venne D., Durivage A. and Vaillancourt J.-P. (2011). Description of 44 biosecurity errors while entering and exiting poultry barns based on video surveillance in Quebec, Canada. *Prev. Vet. Med.* 100, 193–199.
- Conan, A., Goutard, F. L., Sorn, S. and Vong, S. (2012). Biosecurity measures for backyard poultry in developing countries: a systematic review. *BMC Veterinary Research* 8(1), 1–10.
- Scott, A. B., Singh, M., Groves, P., Hernandez-Jover, M., Barnes, B., Glass, K., Moloney, B., Black, A. and Toribio, J. A. (2018). Biosecurity practices on Australian commercial layer and meat chicken farms: Performance and perceptions of farmers. PLoS ONE 13(4), e0195582.

Key research centers readers can investigate, for example, for possible collaboration as well as to keep up with research trends:

1. Prof. Dr. Jeroen Dewulf: https://biocheck.ugent.be/en:
   Veterinary Epidemiology Unit, Department of Reproduction, Obstetrics and Herd Health.
   Faculty of Veterinary Medicine; Ghent University, Belgium:
   https://www.ugent.be/di/vvb/en/research/research-epidemiology.
2. Prof. Jean-Pierre Vaillancourt: Department of Clinical Sciences; Faculty of Veterinary Medicine, Université de Montréal, Canada. Jean-pierre.vaillancourt@umontreal .ca.
3. Jean-Luc Guérin. Poultry Biosecurity Chair, École Nationale Vétérinaire de Toulouse; jl.guerin@envt.fr:
   http://www.envt.fr/content/recherche-publication-de-la-chaire-de -biosécurité-aviaire.
4. Alberto Oscar Allepuz Palau: Alberto.Allepuz@uab.cat:
   Centre de Recerca en Sanitat Animal IRTA-CReSA; University of Barcelona, Spain.
5. Armin R. W. Elbers; armin.elbers@wur.nl:
   Department of Epidemiology, Bioinformatics and Animal Models, Wageningen Bioveterinary Research, The Netherlands.

14 References

Allen, V. and Newell, D. (2005). *Evidence for the Effectiveness of Biosecurity to Exclude Campylobacter from Poultry Flocks.* Commissioned Project MS0004. Food Standards Agency Report. Available at: https://pdfs.semanticscholar.org/4236/1dd1fb80d51742c54629ee46cb315ef936a3.pdf?_ga=2.63452588.287620099.1573885102-1033270699.1521297455 (accessed 22 March 2021).

Amass, S. F. (1999). Biosecurity considerations for pork production units, *Swine Health and Production* 7(5), 12.

Amass, S. F., Vyverberg, B. D., Ragland, D., Dowell, C. A., Anderson, C. D., Stover, J. H. and Beaudry, D. J. (2000). Evaluating the efficacy of boot baths in biosecurity protocols, *Journal of Swine Health and Production* 8(4), 169–173.

Amass, S. F., Arighi, M., Kinyon, J. M., Hoffman, L. J., Schneider, J. L. and Draper, D. K. (2006). Effectiveness of using a mat filled with a peroxygen disinfectant to minimize shoe sole contamination in a veterinary hospital, *Journal of the American Veterinary Medical Association* 228(9), 1391–1396.

Anon. (2018). National on-farm avian biosecurity standards. Available at: https://inspection.canada.ca/animal-health/terrestrial-animals/biosecurity/standards-and-principles/national-avian-on-farm-biosecurity-standard/eng/1528732756921/1528732872665?chap=0 (accessed 18 April 2022).

Arsenault, J., Letellier, A., Quessy, S., Normand, V. and Boulianne, M. (2007). Prevalence and risk factors for *Salmonella* spp. and *Campylobacter* spp. caecal colonization in broiler chicken and turkey flocks slaughtered in Quebec, Canada, *Preventive Veterinary Medicine* 81(4), 250–264.

Axtell, R. C. (1999). Poultry integrated pest management: status and future, *Integrated Pest Management Reviews* 4(1), 53–73.

Bandura, A. and Cervone, D. (1983). Self-evaluative and self-efficacy mechanisms governing the motivational effects of goal systems, *Journal of Personality and Social Psychology* 45(5), 1017–1028. doi: 10.1037/0022-3514.45.5.1017.

Barcelo, M. and Marco, E. (1998). On farm biosecurity. *Proceedings of the 15th Intl Pig. Veterinary Society Congress.* Nottingham University Press, Birmingham, England.

Barrington, G. M., Allen, A. J., Parish, S. M. and Tibary, A. (2006). Biosecurity and biocontainment in alpaca operations, *Small Ruminant Research* 61(2), 217–225. doi: 10.1016/j.smallrumres.2005.07.011.

Bates, C., Hiett, K. L. and Stern, N. J. (2004). Relationship of campylobacter isolated from poultry and from darkling beetles in New Zealand, *Avian Diseases* 48(1), 138–147.

Battersby, T., Walsh, D., Whyte, P. and Bolton, D. (2017). Evaluating and improving terminal hygiene practices on broiler farms to prevent Campylobacter cross-contamination between flocks, *Food Microbiology* 64, 1–6. doi: 10.1016/j.fm.2016.11.018.

Bennett, B. (2017). The importance of biosecurity in the modern day hatchery, *International Hatchery Practice* 31, 21–23.

Bestman, M., de Jong, W., Wagenaar, J. and Weerts, T. (2018). Presence of avian influenza risk birds in and around poultry free-range areas in relation to range vegetation and openness of surrounding landscape, *Agroforestry Systems* 92(4), 1001–1008. doi: 10.1007/s10457-017-0117-2.

Blondel, V., Huard, G., Vaillancourt, J. P. and Racicot, M. (2018). Base du nettoyage et de la désinfection dans les exploitations agricoles. Available at: https://www.agrireseau.net/documents/98011/bases-du-nettoyage-et-de-la-desinfection-dans-les-exploitations-agricoles (accessed 7 October 2021).

Boender, G. J., Meester, R., Gies, E. and De Jong, M. C. (2007). The local threshold for geographical spread of infectious diseases between farms, *Preventive Veterinary Medicine* 82(1–2), 90–101. doi: 10.1016/j.prevetmed.2007.05.016.

Böhm, R. (1998). Disinfection and hygiene in the veterinary field and disinfection of animal houses and transport vehicles, *International Biodeterioration and Biodegradation* 41(3-4), 217-224.

Boklund, A., Alban, L., Mortensen, S. and Houe, H. (2004). Biosecurity in 116 Danish fattening swineherds: descriptive results and factor analysis, *Preventive Veterinary Medicine* 66(1-4), 49-62. doi: 10.1016/j.prevetmed.2004.08.004.

Bonhotal, J., Schwarz, M. and Rynk, R. (2014). Composting animal mortalities. Cornell Waste Management Institute, 1-23. Available at: https://hdl.handle.net/1813/37369.

Bouwstra, R., Gonzales, J. L., de Wit, S., Stahl, J., Fouchier, R. A. M. and Elbers, A. R. W. (2017). Risk for low pathogenicity avian influenza virus on poultry farms, the Netherlands, 2007-2013, *Emerging Infectious Diseases* 23(9), 1510-1516. doi: 10.3201/eid2309.170276.

Bradley, F. A. (2007). Biosecurity: educational programs, *Journal of Applied Poultry Research* 16(1), 77-81. doi: 10.1093/japr/16.1.77.

Brglez, B. (2003). *Disposal of Poultry Carcasses in Catastrophic Avian Influenza Outbreaks*. doi: 10.17615/vxd0-mv88.

Brody, S. N. (1974). The disease of the soul: leprosy in medieval literature (1st ed.). Ithaca: Cornell University Press.

Buckalew, L. W. and Sallis, R. E. (1986). Patient compliance and medication perception, *Journal of Clinical Psychology* 42(1), 49-53. doi: 10.1002/1097-4679(198601)42:1<49::AID-JCLP227 0420107>3.0.CO;2-F.

Butcher, G. D. and Miles, R. D. (1995). *Minimizing Microbial Contamination in Feed Mills Producing Poultry Feed*. Veterinary Medicine-Large Animal Clinical Sciences Department, Florida Cooperative Extension Service, Institute of Food and Agricultural Sciences and University of Florida. Available at: http://www.nutritime.com.br/arquivos_internos/artigos/Artigo78 _VM054002.pdf.

Casal, J., De Manuel, A., Mateu, E. and Martín, M. (2007). Biosecurity measures on swine farms in Spain: perceptions by farmers and their relationship to current on-farm measures, *Preventive Veterinary Medicine* 82(1-2), 138-150. doi: 10.1016/j.prevetmed.2007.05.012.

Cassar, J. R., Bright, L. M., Patterson, P. H., Mills, E. W. and Demirci, A. (2020). The efficacy of pulsed ultraviolet light processing for table and hatching eggs, *Poultry Science* 100(3), 100923.

Cerf, O., Carpentier, B. and Sanders, P. (2010). Tests for determining in-use concentrations of antibiotics and disinfectants are based on entirely different concepts: "Resistance" has different meanings, *International Journal of Food Microbiology* 136(3), 247-254.

Chaber, A. L. and Saegerman, C. (2017). Biosecurity measures applied in the United Arab Emirates – a comparative study Between livestock and wildlife sectors, *Transboundary and Emerging Diseases* 64(4), 1184-1190. doi: 10.1111/tbed.12488.

Chaudhry, M., Ahmad, M., Rashid, H. B., Sultan, B., Chaudhry, H. R., Riaz, A. and Shaheen, M. S. (2017). Prospective study of avian influenza H9 infection in commercial poultry farms of Punjab Province and Islamabad Capital Territory, Pakistan, *Tropical Animal Health and Production* 49(1), 213-220. doi: 10.1007/s11250-016-1159-6. PMID: 27761776; PMCID: PMC7088531.

Chen, S. J., Hung, M. C., Huang, K. L. and Hwang, W. I. (2004). Emission of heavy metals from animal carcass incinerators in Taiwan, *Chemosphere* 55(9), 1197-1205.

Chin, R. P., García, M., Corsiglia, C., Riblet, S., Crespo, R., Shivaprasad, H. L., Rodríguez-Avila, A., Woolcock, P. R. and França, M. (2009). Intervention strategies for laryngotracheitis: impact of extended downtime and enhanced biosecurity auditing, *Avian Diseases* 53(4), 574-577. doi: 10.1637/8873-041309-Reg.1.

Cochrane, R. A. (2016). Feed mill biosecurity plans: a systematic approach to prevent biological pathogens in swine feed, *Journal of Swine Health and Production* 24(3), 154-164.

Colombo, C., Giger, H., Grote, J., Deplazes, C., Pletscher, W., Lüthi, R. and Ruef, C. (2002). Impact of teaching interventions on nurse compliance with hand disinfection, *Journal of Hospital Infection* 51(1), 69-72. doi: 10.1053/jhin.2002.1198.

Conly, J. M., Hill, S., Ross, J., Lertzman, J. and Louie, T. J. (1989). Handwashing practices in an intensive care unit: the effects of an educational program and its relationship to infection rates, *American Journal of Infection Control* 17(6), 330-339. doi: 10.1016/0196-6553(89)90002-3.

Corrigan, R. M. (2006). Overview of rodent control for commercial pork operation. Available at: https://porkgateway.org/resource/an-overview-of-rodent-control-for-commercial-pork-production-operations/.

Course, C. E., Boerlin, P., Slavic, D., Vaillancourt, J. P. and Guerin, M. T. (2021). Factors associated with *Salmonella enterica* and *Escherichia coli* during downtime in commercial broiler chicken barns in Ontario, *Poultry Science* 100(5), 101065.

Crippen, T. L. and Sheffield, C. (2006). External surface disinfection of the lesser mealworm (Coleoptera: Tenebrionidae), *Journal of Medical Entomology* 43(5), 916-923.

Curtis, P. E., Ollerhead, G. E. and Ellis, C. E. (1980). *Pasteurella multocida* infection of poultry farm rats, *Veterinary Record* 107(14), 326-327.

Curtis, P. E. and Ollerhead, G. E. (1982). *Pasteurella multocida* infection of cats on poultry farms, *Veterinary Record* 110(1), 13-14.

Davies, R. H. and Wray, C. (1995). Observations on disinfection regimens used on *Salmonella enteritidis* infected poultry units, *Poultry Science* 74(4), 638-647.

Davison, S. A., Dunn, P. A., Henzler, D. J., Knabel, S. J., Patterson, P. H. and Schwartz, J. H. (1997). *Preharvest HACCP in the Table Egg Industry*. PA State: The Pennsylvania State University, 1-36.

Dee, S., Deen, J., Rossow, K., Wiese, C., Otake, S., Joo, H. S. and Pijoan, C. (2002). Mechanical transmission of porcine reproductive and respiratory syndrome virus throughout a coordinated sequence of events during cold weather, *Canadian Journal of Veterinary Research* 66(4), 232-239.

Dee, S., Deen, J. and Pijoan, C. (2004). Evaluation of 4 intervention strategies to prevent the mechanical transmission of porcine reproductive and respiratory syndrome virus, *Canadian Journal of Veterinary Research* 68(1), 19-26.

Delabbio, J. L., Johnson, G. R., Murphy, B. R., Hallerman, E., Woart, A. and McMullin, S. L. (2005). Fish disease and biosecurity: attitudes, beliefs, and perceptions of managers and owners of commercial finfish recirculating facilities in the United States and Canada, *Journal of Aquatic Animal Health* 17(2), 153-159. doi: 10.1577/H04-005.1.

Delpont, M., Racicot, M., Durivage, A., Fornili, L., Guerin, J. L., Vaillancourt, J. P. and Paul, M. C. (2020). Determinants of biosecurity practices in French duck farms after a H5N8 Highly Pathogenic Avian Influenza epidemic: the effect of farmer knowledge, attitudes and personality traits, *Transboundary and Emerging Diseases* 68(1), 51-61.

Dorea, F. C., Berghaus, R., Hofacre, C. and Cole, D. J. (2010). Survey of biosecurity protocols and practices adopted by growers on commercial poultry farms in Georgia, USA, *Avian Diseases* 54(3), 1007-1015.

Dunowska, M., Morley, P. S., Patterson, G., Hyatt, D. R. and Van Metre, D. C. (2006). Evaluation of the efficacy of a peroxygen disinfectant-filled footmat for reduction of bacterial load on footwear in a large animal hospital setting, *Journal of the American Veterinary Medical Association* 228(12), 1935-1939.

Duvauchelle, A., Huneau-Salaün, A., Balaine, L., Rose, N. and Michel, V. (2013). Risk factors for the introduction of avian influenza virus in breeder duckflocksduring the first 24weeks of laying, *Avian Pathology* 42(5), 447-456.doi: 10.1080/03079457.2013.823145.

East, I., Kite, V., Daniels, P. and Garner, G. (2006). A cross-sectional survey of Australian chicken farms to identify risk factors associated with seropositivity to Newcastle-disease virus, *Preventive Veterinary Medicine* 77(3-4), 199-214. doi: 10.1016/j.prevetmed.2006.07.004.

Ebeling, W. (1975). *Urban Entomology*. Los Angeles: University of California.

Elfadil, A. A., Vaillancourt, J. P. and Meek, A. H. (1996). Management risk factors associated with cellulitis in broiler chickens in Southern Ontario: a retrospective study, *Avian Diseases* 40(3), 699-706.

Ellis, D. B. (2001). Carcass disposal issues in recent disasters, accepted methods, and suggested plan to mitigate future events. Available at: https://digital.library.txstate.edu/handle/10877/3502 (accessed 18 April 2022).

England, J. J. (2002). Biosecurity: safeguarding your veterinarian:client:patient relationship, *Veterinary Clinics of North America. Food Animal Practice* 18(3), 373-8, v. doi: 10.1016/ S0749-0720(02)00033-6.

Ernst, A. R. (2004). *Hatching Egg Sanitation: The Key Step in Successful Storage and Production.* University of California, Division of Agriculture and National Resources: Publication 8120.

Fasina, F. O., Lazarus, D. D., Spencer, B. T., Makinde, A. A. and Bastos, A. D. S. (2012). Cost implications of African swine fever in smallholder farrow-to-finish units: economic benefits of disease prevention through biosecurity, *Transboundary and Emerging Diseases* 59, 244-255.

Fernandez, D., et al. (1994). Farm location as a determinant to production performance in turkeys. Annual Meeting of the American Association of Avian Pathologists, Annual Convention AVMA, San Francisco.

Fraser, R. W., Williams, N. T., Powell, L. F. and Cook, A. J. C. (2010). Reducing campylobacter and salmonella infection: two studies of the economic cost and attitude to adoption of on-farm biosecurity measures, *Zoonoses and Public Health* 57(7-8), e109-e115.

Gelaude, P., Schlepers, M., Verlinden, M., Laanen, M. and Dewulf, J. (2014). Biocheck. UGent: a quantitative tool to measure biosecurity at broiler farms and the relationship with technical performances and antimicrobial use, *Poultry Science* 93(11), 2740-2751.

Georgiades, E., Fraser, R. and Jones, B. (2016). Options to strengthen on-farm biosecurity management for commercial and non-commercial aquaculture. Aquaculture Unit. Technical Paper No: 2016/47.

Gifford, D. H., Shane, S. M., Hugh-Jones, M. and Weigler, B. J. (1987). Evaluation of biosecurity in broiler breeders, *Avian Diseases* 31(2), 339-344. doi: 10.2307/1590882.

Glanville, T. D., Ahn, H. K., Richard, T. L., Harmon, J. D., Reynolds, D. L. and Akinc, S. (2006). Environmental impacts of emergency livestock mortality composting-leachate release and soil contamination. 2006 ASAE Annual Meeting (p. 1). American Society of Agricultural and Biological Engineers.

Graham, J. P., Price, L. B., Evans, S. L., Graczyk, T. K. and Silbergeld, E. K. (2009). Antibiotic resistant enterococci and staphylococci isolated from flies collected near confined poultry feeding operations, *Science of the Total Environment* 407(8), 2701-2710.

Guinat, C., Comin, A., Kratzer, G., Durand, B., Delesalle, L., Delpont, M., Guérin, J. L. and Paul, M. C. (2020). Biosecurity risk factors for highly pathogenic avian influenza (H5N8) virus infection in duck farms, France, *Transboundary and Emerging Diseases* 67(6), 2961-2970. doi: 10.1111/ tbed.13672.

Gunn, G. J., Heffernan, C., Hall, M., McLeod, A. and Hovi, M. (2008). Measuring and comparing constraints to improved biosecurity amongst GB farmers, veterinarians and the auxiliary industries, *Preventive Veterinary Medicine* 84(3-4), 310-323. doi: 10.1016/j. prevetmed.2007.12.003.

Hald, B., Skovgard, H., Bang, D. D., Pedersen, K., Dybdahl, J., Jespersen, J. B. and Madsen, M. (2004). Flies and campylobacter infection of broiler flocks, *Emerging Infectious Diseases* 10(8), 1490-1492.

Halvorson, D. A. (2002). The control of H5 or H7 mildly pathogenic avian influenza: a role for inactivated vaccine. *Avian Pathology* 31(1):5-12. doi: 10.1080/03079450120106570. PMID: 12430550.

Hauck, R., Crossley, B., Rejmanek, D., Zhou, H. and Gallardo, R. A. (2017). Persistence of highly pathogenic and low pathogenic avian influenza viruses in footbaths and poultry manure. *Avian Diseases* 61(1), 64-69. doi: 10.1637/11495-091916-Reg.

Haynes, R. B., Taylor, D. W. and Sackett, D. L. (Eds) (1979). *Compliance in Health Care.* Baltimore: Johns Hopkins University Press.

Henzler, D. J. and Opitz, H. M. (1992). The role of mice in the epizootiology of *Salmonella enteritidis* infection on chicken layer farms, *Avian Diseases* 36(3), 625-631.

Jardine, C. G. and Hrudey, S. E. (1997). Mixed messages in risk communication, *Risk Analysis* 17(4), 489-498. doi: 10.1111/j.1539-6924.1997.tb00889.x.

Jeong, J., Kang, H. M., Lee, E. K., Song, B. M., Kwon, Y. K., Kim, H. R., Choi, K. S., Kim, J. Y., Lee, H. J., Moon, O. K., Jeong, W., Choi, J., Baek, J. H., Joo, Y. S., Park, Y. H., Lee, H. S. and Lee, Y. J. (2014). Highly pathogenic avian influenza virus (H5N8) in domestic poultry and its relationship with migratory birds in South Korea during 2014, *Veterinary Microbiology* 173(3-4), 249–257.

Johnson, Y. J., Gedamu, N., Colby, M. M., Myint, M. S., Steele, S. E., Salem, M. and Tablante, N. L. (2005). Wind-borne transmission of infectious laryngotracheitis between commercial poultry operations, *International Journal of Poultry Science* 4(5), 263–267. doi: 10.3923/ijps.2005.263.267.

Jones, F. T. (2011). A review of practical Salmonella control measures in animal feed, *Journal of Applied Poultry Research* 20(1), 102–113.

Pinto, C. J. and Urcelay, V. S. (2003). Biosecurity practices on intensive pig production systems in Chile, *Preventive Veterinary Medicine* 59(3), 139–145. doi: 10.1016/S0167-5877(03)00074-6.

Kalbasi, A., Mukhtar, S., Hawkins, S. E. and Auvermann, B. W. (2005). Carcass composting for management of farm mortalities: a review, *Compost Science and Utilization* 13(3), 180–193.

Keener, H. M., Elwell, D. L. and Monnin, M. J. (2000). Procedures and equations for sizing of structures and windrows for composting animal mortalities, *Applied Engineering in Agriculture* 16(6), 681–692.

Kim, J. H. and Kim, K. S. (2010). Hatchery hygiene evaluation by microbiological examination of hatchery samples, *Poultry Science* 89(7), 1389–1398.

Kim, W. H., An, J. U., Kim, J., Moon, O. K., Bae, S. H., Bender, J. B. and Cho, S. (2018). Risk factors associated with highly pathogenic avian influenza subtype H5N8 outbreaks on broiler duck farms in South Korea, *Transboundary and Emerging Diseases* 65(5), 1329–1338.

King, M. A., Seekins, B., Hutchinson, M. and MacDonald, G. (2005). Observations of static pile composting of large animal carcasses using different media. Symposium on Composting Animal Mortalities and Slaughterhouse Residuals in South Portland, ME.

Korbel, R., Gerlach, H., Bisgaard, M. and Hafez, H. M. (1992). Further investigations on *Pasteurella multocida* infections in feral birds injured by cats, *Zentralblatt Fur Veterinarmedizin. Reihe B. Journal of Veterinary Medicine. Series B* 39(1), 10–18.

Kuney, D. R. and Jeffrey, J. S. (2002). Cleaning and disinfecting poultry facilities. In: Bell, D. D., Weaver, W. D. (Eds) *Commercial Chicken Meat and Egg Production*, 557–564. Boston: Springer. doi: 10.1007/978-1-4615-0811-3_29.

Langsrud, S., Møretrø, T. and Sundheim, G. (2003). Characterization of *Serratia marcescens* surviving in disinfecting footbaths, *Journal of Applied Microbiology* 95(1), 186–195. doi: 10.1046/j.1365-2672.2003.01968.x.

Larsen, A. F. (2009). Semi-subsistence producers and biosecurity in the Slovenian alps, *Sociologia Ruralis* 49(4), 330–343. doi: 10.1111/j.1467-9523.2009.00481.x.

Lee, H. J., Jeong, J. Y., Jeong, O. M., Youn, S. Y., Kim, J. H., Kim, D. W., Yoon, J. U., Kwon, Y. K. and Kang, M. S. (2020). Impact of Dermanyssus gallinae infestation on persistent outbreaks of fowl typhoid in commercial layer chicken farms, *Poultry Science* 99(12), 6533–6541.

Li, X., Bethune, L. A., Jia, Y., Lovell, R. A., Proescholdt, T. A., Benz, S. A., Schell, T. C., Kaplan, G. and McChesney, D. G. (2012). Surveillance of *Salmonella* prevalence in animal feeds and characterization of the *Salmonella* isolates by serotyping and antimicrobial susceptibility, *Foodborne Pathogens and Disease* 9(8), 692–698. doi: 10.1089/fpd.2011.1083.

Losinger, W. C., Bush, E. J., Hill, G. W., Smith, M. A., Garber, L. P., Rodriguez, J. M. and Kane, G. (1998). Design and implementation of the United States National Animal Health Monitoring System 1995 National Swine Study, *Preventive Veterinary Medicine* 34(2-3), 147–159. doi: 10.1016/s0167-5877(97)00076-7.

Lotz, J. M. (1997). Special topic review: viruses, biosecurity and specific pathogen-free stocks in shrimp aquaculture, *World Journal of Microbiology and Biotechnology* 13, 9.

Mauldin, J. M. (1983). *Hatchery and Breeder Flock Sanitation Guide*. Bulletin-Cooperative Extension Service. Available at: agris.fao.org.

Mauldin, J. M. (2002). Hatchery planning, design, and construction. In: Bell, D. D., Weaver, W. D. and North, M. O. (Eds) *Commercial Chicken Meat and Egg Production*, 661–683. Boston: Springer.

Mauldin, J. M. and MacKinnon, I. R. (2009). Hatchery ventilation and environmental control, *Avian Biology Research* 2(1–2), 87–91.

McQuiston, J. H., Garber, L. P., Porter-Spalding, B. A., Hahn, J. W., Pierson, F. W., Wainwright, S. H., Senne, D. A., Brignole, T. J., Akey, B. L. and Holt, T. J. (2005). Evaluation of risk factors for the spread of low pathogenicity H7N2 avian influenza virus among commercial poultry farms, *Journal of the American Veterinary Medical Association* 226(5), 767–772.

Millman, C., Christley, R., Rigby, D., Dennis, D., O'Brien, S. J. and Williams, N. (2017). "Catch 22": biosecurity awareness, interpretation and practice amongst poultry catchers, *Preventive Veterinary Medicine* 141, 22–32.

Moore, D. A., Merryman, M. L., Hartman, M. L. and Klingborg, D. J. (2008). Comparison of published recommendations regarding biosecurity practices for various production animal species and classes, *Journal of the American Veterinary Medical Association* 233(2), 249–256. doi: 10.2460/javma.233.2.249.

Morley, P. S., Morris, S. N., Hyatt, D. R.and Van Metre, D. C. (2005). Evaluation of the efficacy of disinfectant footbaths as used in veterinary hospitals, *Journal of the American Veterinary Medical Association* 226(12), 2053–2058.

Moro, C. V., De Luna, C. J., Tod, A., Guy, J. H., Sparagano, O. A. and Zenner, L. (2009). The poultry red mite (Dermanyssus gallinae): a potential vector of pathogenic agents. In: Sparagano, O. A. (Ed.) *Control of Poultry Mites (Dermanyssus)*, 93–104. Dordrecht: Springer.

Morris, M. P. (1995). Economic considerations in prevention and control of poultry disease. In: *Biosecurity in the Poultry Industry*, Shane, S. M., Halvorson, D., Hill, D., Villegas, P. and Wages, D. (Eds), 4–16. Kennett Square: American Association of Avian Pathologists.

Mukhtar, S., Kalbasi, A. and Ahmed, A. (2004). *Composting: A Comprehensive Review*. Kansas State University, National Agricultural Biosecurity Center. Available at: https://krex.k-state.edu/dspace/bitstream/handle/2097/662/Chapter3.pdf?sequence=16&isAllowed=y.

Nespeca, R., Vaillancourt, J. P. and Morrow, W. E. (1997). Validation of a poultry biosecurity survey, *Preventive Veterinary Medicine* 31(1–2), 73–86.

Niemi, J. K., Sahlström, L., Kyyrö, J., Lyytikäinen, T. and Sinisalo, A. (2016). Farm characteristics and perceptions regarding costs contribute to the adoption of biosecurity in Finnish pig and cattle farms, *Review of Agricultural, Food and Environmental Studies* 97(4), 215–223. doi: 10.1007/s41130-016-0022-5.

Nishiguchi, A., Kobayashi, S., Yamamoto, T., Ouchi, Y., Sugizaki, T. and Tsutsui, T. (2007). Risk factors for the introduction of avian influenza virus into commercial layer chicken farms During the outbreaks caused by a low-pathogenic H5N2 virus in Japan in 2005, *Zoonoses and Public Health* 54(9–10), 337–343. doi: 10.1111/j.1863-2378.2007.01074.x.

O'Boyle, C. A., Henly, S. J. and Larson, E. (2001). Understanding adherence to hand hygiene recommendations: the theory of planned behavior, *American Journal of Infection Control* 29(6), 352–360. doi: 10.1067/mic.2001.18405.

Ojewole, A. O. (2011). Biblical mandates for sustainable sanitation, *Millennium Development Goals (MdGS) as Instrument for Development in Africa*, 598–613. Available at: https://d1wqtxts1xzle7.cloudfront.net/49544347/How_green_are_hotels_in_Accra_Environmen20161012-27178-u5ze10-with-cover-page-v2.pdf?Expires=1650324230&Signature=JFBCUcC8IbV5l4WMUpjOQBc-xpAsgUu0WZKcq-ZnnZ7pq2Ua~ccfwUSN-6YyQkCAoTlxr1h0qks8EVWkpaA3m315SoPjCoFUEGFka9-3QYp9wC7rRDRSAX7a~g9EgMH253tYuJW5Z67hMcLH5mNQwdLwmCFQmwAaLzzvk4c-xh5V6l-1Y576Klzv8SkCS0giTlCCqRfmb-Tv5hmh9JoKz9K0C7qRmHu0XlmMf31j6NecmUm-Xhh0PwelFFxkoGL~BzwkZ5TQ~Upyb7RefKAOjze65X2XdCCuri-14XVRvBK~n-ITgh3FoX-73onLAx0QfXr7ACdxvSgotDmDxYvmBQ__&Key-Pair-Id=APKAJLOHF5GGSLRBV4ZA#page=609.

Owen, R. L. and Lawlor, J. (2012). A novel approach to foot dipping. Available at: https://fr.slideserve.com/bernad/a-novel-approach-to-foot-dipping.

Pagès-Manté, A., Torrents, D., Maldonado, J. and Saubi, N. (2004). Dogs as potential carriers of infectious bursal disease virus, *Avian Pathology* 33(2), 205-209.

Pandolfi, F., Edwards, S. A., Maes, D. and Kyriazakis, I. (2018). Connecting different data sources to assess the interconnections between biosecurity, health, welfare, and performance in commercial pig farms in Great Britain, *Frontiers in Veterinary Science* 5, 41. doi: 10.3389/fvets.2018.00041.

Payne, J. B., Kroger, E. C. and Watkins, S. E. (2005). Evaluation of disinfectant efficacy when applied to the floor of poultry grow-out facilities, *Journal of Applied Poultry Research* 14(2), 322-329.

Pervin, L. A. and John, O. P. (2001). *Personality: Theory and Research* (8th edn.). New York: John Wiley & Sons Inc.

Pudenz, C. C., Schulz, L. L. and Tonsor, G. T. (2019). Adoption of secure pork supply plan biosecurity by US Swine producers, *Frontiers in Veterinary Science* 6, 146.

Rabie, A. J., McLaren, I. M., Breslin, M. F., Sayers, R. and Davies, R. H. (2015). Assessment of anti-Salmonella activity of boot dip samples, *Avian Pathology* 44(2), 129-134.

Racicot, M., Venne, D., Durivage, A. and Vaillancourt, J. P. (2011). Description of 44 biosecurity errors while entering and exiting poultry barns based on video surveillance in Quebec, Canada, *Preventive Veterinary Medicine* 100(3-4), 193-199. doi: 10.1016/j.prevetmed.2011.04.011.

Racicot, M., Venne, D., Durivage, A. and Vaillancourt, J. P. (2012a). Evaluation of strategies to enhance biosecurity compliance on poultry farms in Québec: effect of audits and cameras, *Preventive Veterinary Medicine* 103(2-3), 208-218. doi: 10.1016/j.prevetmed.2011.08.004.

Racicot, M., Venne, D., Durivage, A. and Vaillancourt, J. P. (2012b). Evaluation of the relationship between personality traits, experience, education and biosecurity compliance on poultry farms in Québec, Canada, *Preventive Veterinary Medicine* 103(2-3), 201-207. doi: 10.1016/j.prevetmed.2011.08.011.

Racicot, M., Comeau, G., Tremblay, A., Quessy, S., Cereno, T., Charron-Langlois, M., Venne, D., Hébert, G., Vaillancourt, J. P., Fravalo, P., Ouckama, R., Mitevski, D., Guerin, M. T., Agunos, A., DeWinter, L., Catford, A., Mackay, A. and Gaucher, M. L. (2020). Identification and selection of food safety-related risk factors to be included in the Canadian Food Inspection Agency's Establishment-based Risk Assessment model for Hatcheries, *Zoonoses and Public Health* 67(1), 14-24.

Refrégier-Petton, J., Rose, N., Denis, M. and Salvat, G. (2001). Risk factors for Campylobacter spp. contamination in French broiler-chicken flocks at the end of the rearing period, *Preventive Veterinary Medicine* 50(1-2), 89-100. doi: 10.1016/s0167-5877(01)00220-3.

Ribbens, S., Dewulf, J., Koenen, F., Mintiens, K., De Sadeleer, L., de Kruif, A. and Maes, D. (2008). A survey on biosecurity and management practices in Belgian pig herds, *Preventive Veterinary Medicine* 83(3-4), 228-241. doi: 10.1016/j.prevetmed.2007.07.009.

Ricke, S. C., Richardson, K. and Dittoe, D. K. (2019). Formaldehydes in feed and their potential interaction with the poultry gastrointestinal tract microbial community-a review, *Frontiers in Veterinary Science* 6, 188. doi: 10.3389/fvets.2019.00188.

Roche, A. J., Cox, N. A., Richardson, L. J., Buhr, R. J., Cason, J. A., Fairchild, B. D. and Hinkle, N. C. (2009). Transmission of *Salmonella* to broilers by contaminated larval and adult lesser mealworms, *Alphitobius diaperinus* (Coleoptera: Tenebrionidae), *Poultry Science* 88(1), 44-48.

Rodgers, J. D., McCullagh, J. J., McNamee, P. T., Smyth, J. A. and Ball, H. J. (2001). An investigation into the efficacy of hatchery disinfectants against strains of Staphylococcus aureus associated with the poultry industry, *Veterinary Microbiology* 82(2), 131-140.

Rodrigues da Costa, M., Gasa, J., Calderón Díaz, J. A., Postma, M., Dewulf, J., McCutcheon, G. and Manzanilla, E. G. (2019). Using the biocheck. UGent™ scoring tool in Irish farrow-to-finish pig farms: assessing biosecurity and its relation to productive performance, *Porcine Health Management* 5, 4.

Rojo-Gimeno, C., Postma, M., Dewulf, J., Hogeveen, H., Lauwers, L. and Wauters, E. (2016). Farm-economic analysis of reducing antimicrobial use whilst adopting improved management strategies on farrow-to-finish pig farms, *Preventive Veterinary Medicine* 129, 74–87. doi: 10.1016/j.prevetmed.2016.05.001.

Rose, N., Beaudeau, F., Drouin, P., Toux, J. Y., Rose, V. and Colin, P. (2000). Risk factors for Salmonella persistence after cleansing and disinfection in French broiler-chicken houses, *Preventive Veterinary Medicine* 44(1–2), 9–20. doi: 10.1016/S0167-5877(00)00100-8.

Roy, D. N. and Brown, A. W. A. (1954). *Entomology : (Medical & Veterinary) Including Insecticides & Insect & Rat Control*. Calcutta: Excelsior Press.

Sambeek, F. V., McMurray, B. L. and Page, R. K. (1995). Incidence of *Pasteurella multocida* in poultry house cats used for rodent control programs, *Avian Diseases* 39(1), 145–146.

Sanchez, J., Stryhn, H., Flensburg, M., Ersbøll, A. K. and Dohoo, I. (2005). Temporal and spatial analysis of the 1999 outbreak of acute clinical infectious bursal disease in broiler flocks in Denmark, *Preventive Veterinary Medicine* 71(3–4), 209–223. doi: 10.1016/j.prevetmed.2005.07.006.

Sander, J. E., Warbington, M. C. and Myers, L. M. (2002). Selected methods of animal carcass disposal, *Journal of the American Veterinary Medical Association* 220(7), 1003–1005.

Sanderson, M. W., Dargatz, D. A. and Garry, F. B. (2000). Biosecurity practices of beef cow-calf producers, *Journal of the American Veterinary Medical Association* 217(2), 185–189. doi: 10.2460/javma.2000.217.185.

Sawabe, K., Hoshino, K., Isawa, H., Sasaki, T., Hayashi, T., Tsuda, Y., Kurahashi, H., Tanabayashi, K., Hotta, A., Saito, T., Yamada, A. and Kobayashi, M. (2006). Detection and isolation of highly pathogenic H5N1 avian influenza A viruses from blow flies collected in the vicinity of an infected poultry farm in Kyoto, Japan, 2004, *American Journal of Tropical Medicine and Hygiene* 75(2), 327–332.

Schoof, H. F. (1959). How far do flies fly and what effect does flight pattern have on their control, *Pest Control* 27(4), 16–24.

Shaw, C. D. and Costain, D. W. (1989). Guidelines for medical audit: seven principles, *BMJ* 299(6697), 498–499.

Smith, T. (1990). Medical audit, *BMJ* 300(6717), 65–65.

Snow, L. C., Davies, R. H., Christiansen, K. H., Carrique-Mas, J. J., Cook, A. J. and Evans, S. J. (2010). Investigation of risk factors for Salmonella on commercial egg-laying farms in Great Britain, 2004–2005, *The Veterinary Record* 166(19), 579–586. doi: 10.1136/vr.b4801.

Springthorpe, S. (2000). La désinfection des surfaces et de l'équipement, *Journal of Cancer Dent. Assoc* 66, 558–560.

Steiner, J. J. (2020). *Disinfection of Hatching Eggs Using Low-Energy Electron Beam* (Doctoral dissertation, University of Zurich).

Strömberg, A., Broström, A., Dahlström, U. and Fridlund, B. (1999). Factors influencing patient compliance with therapeutic regimens in chronic heart failure: a critical incident technique analysis, *Heart and Lung* 28(5), 334–341. doi: 10.1053/hl.1999.v28.a99538.

Tablante, N. L. and Malone, G. W. (2006). Controlling avian influenza through in-house composting of depopulated flocks: sharing Delmarva's experience. Proceedings of the 2006 National Symposium on Carcass Disposal. Available at: https://www.animalmortmgmt.org/wp-content/uploads/2014/04/Controlling-AI-through-in-House-Composting-of-De-Populated-F.pdf.

Thermote, L. (2006). Effective hygiene within the hatchery, *International Hatchery Practice* 20(5), 18–21.

Tibballs, J. (1996). Teaching hospital medical staff to handwash, *The Medical Journal of Australia* 164(7), 395–398.

Timmermans, S. and Berg, M. (1997). Standardization in action: achieving local universality through medical protocols, *Social Studies of Science* 27(2), 273–305.

Toma, B., Vaillancourt, J.-P., Dufour, B., Eliot, M., Moutou, F., Marsh, W., Benet, J.-J., Sanaa, M. and Michel, P. (1999). *Dictionary of Veterinary Epidemiology*. Wiley.

Toma, L., Stott, A. W., Heffernan, C., Ringrose, S. and Gunn, G. J. (2013). Determinants of biosecurity behaviour of British cattle and sheep farmers-a behavioural economics analysis, *Preventive Veterinary Medicine* 108(4), 321–333.

Vaillancourt J-P. (1995). Infectious laryngo-tracheitis in broilers: a case-control study, *Annual Meeting of the American Association of Avian Pathologists*, July 11, 1995, Pittsburgh, Pennsylvania

Vaillancourt, J.-P. and Martinez, A. (2001). Relative importance of biosecurity measures: a delphi study. Annual Meeting of the American Association of Avian Pathologists, 138th Annual Convention AVMA, Boston.

Vaillancourt, J.-P. and Carver, D. K. (1998). Biosecurity: perception is not reality, *Poultry Digest* 57(6), 28-36.

Vaillancourt, J.-P., Delpont, M., Racicot, M., Paul, M. and Guérin, J.-L. (2018). Une perspective régionale de la biosécurité. Le nouveau praticien vétérinaire; 10/n°40156.

Valeeva, N. I., van Asseldonk, M. A. and Backus, G. B. (2011). Perceived risk and strategy efficacy as motivators of risk management strategy adoption to prevent animal diseases in pig farming, *Preventive Veterinary Medicine* 102(4), 284-295.

Van De Giessen, A. W., Tilburg, J. J. H. C., Ritmeester, W. S. and Van Der Plas, J. (1998). Reduction of campylobacter infections in broiler Flocks by application of hygiene measures, *Epidemiology and Infection* 121(1), 57-66.

Vandekerchove, D., De Herdt, P., Laevens, H. and Pasmans, F. (2004). Colibacillosis in caged layer hens: characteristics of the disease and the aetiological agent, *Avian Pathology* 33(2), 117-125. doi: 10.1080/03079450310001642149.

Verhagen, J. H., Fouchier, R. A. M. and Lewis, N. (2021). Highly pathogenic avian influenza viruses at the wild-domestic bird interface in Europe: future directions for research and surveillance, *Viruses* 13(2), 212.

Villa, B. and Velasco, A. (1994). Integrated pest management of the rat Rattus norvegicus in poultry farms, *Veterinaria México* 25(3), 247-249.

Volkova, V., Thornton, D., Hubbard, S. A., Magee, D., Cummings, T., Luna, L., Watson, J. and Wills, R. (2012). Factors associated with introduction of infectious laryngotracheitis virus on broiler farms during a localized outbreak, *Avian Diseases* 56(3), 521-528. doi: 10.1637/10046-122111-Reg.1.

Vroom, V. H. (1994). *Work and Motivation*. San Francisco, CA: John Wiley & Sons.

Watkins, S. and Venne, D. (2015). Water quality. In: *Manuel de pathologie aviaire*, Brugère-Picoux, J., Vaillancourt, J.-P., Bouzouaia, M., Shivaprasad, H. L. and Venne, D. (Eds), 560-569. Paris: AFAS.

Young, I., Rajić, A., Letellier, A., Cox, B., Leslie, M., Sanei, B. and McEwen, S. A. (2010). Knowledge and attitudes toward food safety and use of good production practices among Canadian broiler chicken producers, *Journal of Food Protection* 73(7), 1278-1287. doi: 10.4315/0362-028x-73.7.1278.

Zhang, Y. H., Li, C. S., Liu, C. C. and Chen, K. Z. (2013). Prevention of losses for hog farmers in China: insurance, on-farm biosecurity practices, and vaccination, *Research in Veterinary Science* 95(2), 819-824.

Chapter 5

Alternatives to antibiotics in preventing zoonoses and other pathogens in poultry: prebiotics and related compounds

S. C. Ricke, *University of Arkansas, USA*, A. V. S. Perumalla, *Kerry, USA; and* Navam S. Hettiarachchy, *University of Arkansas, USA*

1 Introduction

While bird performance is economically important to the poultry industry, food-borne pathogens that have been identified with poultry, such as *Salmonella* and *Campylobacter*, continue to be a major issue of concern. Although more is being understood about their abilities to colonize the gastrointestinal (GI) tracts of chickens and the subsequent corresponding interaction with the host, questions still remain (Park et al., 2008; Dunkley et al., 2009; Horrocks et al., 2009; Finstad et al., 2012; Howard et al., 2012; Foley et al., 2011, 2013). For example, there are increasing concerns about the risk of developing cross-resistance and multiple antibiotic resistances in pathogenic bacteria of the host animal. In the last few decades, several antibiotics and chemotherapeutic agents have been used in prophylactic doses in poultry feed to improve the poultry performance and thereby achieve economic benefits (Jones and Ricke, 2003). As a result, there is potential for outbreaks of particular enteric diseases originating from food animal sources such as poultry that would greatly compromise public health treatment efforts

http://dx.doi.org/10.19103/AS.2016.0010.06

due to acquired antibiotic resistance (Jones and Ricke, 2003; Chen et al., 2015). These enteric pathogens could lower the productivity as well as increase bird mortality and associated contamination of poultry products for human consumption. These concerns have challenged the meat animal industry to seek a wide range of feed amendment alternatives to not only maintain live animal production proficiency but also retain quality attributes of the resulting food product (Jones and Ricke, 2003; Hajati and Rezaei, 2010; Alloui et al., 2013; Ricke, 2015).

Among the attributes of health and quality that are of concern to the public as well as the processor is microbial contamination that cannot only be a cause of diminished shelf-life, but, depending on the nature of the contaminant, also a food safety risk. While several postharvest strategies are currently implemented to decontaminate and thus minimize spoilage costs and economic losses caused by food-borne diseases/recalls (Ricke et al., 2005; Sofos et al., 2013), preharvest food safety options remain limited. Preharvest food safety control measures, including vaccination, competitive exclusion cultures and the use of natural antimicrobial feed additives such as plant extract compounds, bacteriophage, bacteriocins and their combinations, just to name a few, have all either been proposed, experimentally tested or in some cases currently implemented as means to limit either the colonization and/or carriage of food-borne pathogens such as *Salmonella* associated with poultry (Fuller, 1989; Joerger, 2003; Patterson and Burkholder, 2003; Rastall, 2004; Ricke and Pillai, 1999; Revolledo et al., 2006; Vandeplas et al., 2010; Hume, 2011; Callaway and Ricke, 2012; Ricke et al., 2012a; O'Bryan et al., 2015; Ricke, 2015; Rivera Calo et al., 2015). A number of competitive exclusion cultures, also known as probiotics, have been developed over the years for poultry, some of which consist of a complex consortia of microorganisms, while others rely only on a single organism such as *Bacillus subtilis* (Stavric, 1992; Nisbet et al., 1994; 1996a,b; Ricke and Pillai, 1999; Nisbet, 2002; Siragusa and Ricke, 2012; Ricke and Saengkerdsub, 2015).

Prebiotics were originally defined by Gibson and Roberfroid (1995) as 'a non-digestible food ingredient that beneficially affects the host by selectively stimulating the growth and/or activity of one or a limited number of GI microflora', but that definition has been further refined as more has become understood about the interaction between the gut microbiome and dietary components (Gibson and Roberfroid, 1995; Wang, 2009; Bird et al., 2010; Hutkins et al., 2016). More simply put, prebiotics can generally be considered as a class of compounds that specifically provide substrates (energy) for the beneficial GI microbiome (Hajati and Rezaei, 2010). Prebiotics are non-digestible ingredients; passing through the small intestine to the lower gut where they undergo hydrolysis and subsequent fermentation by colonic and/or cecal bacteria to produce short-chain fatty acids (SCFAs), namely acetate, butyrate and propionate that, in turn, can be absorbed by the host (Demigné et al., 1986; Józefiak et al., 2004; Wang, 2009; Bird et al., 2010).

Prebiotics used in poultry nutrition for preharvest safety include non-digestible carbohydrates such as fructooligosaccharide (FOS) products (oligofructose, inulin-type fructans found in many vegetables), *trans*-galactooligosaccharides (TOS), glucooligosaccharides, glycooligosaccharides, lactulose, lactitol, maltooligosaccharides, soy-oligosaccharides, xylo-oligosaccharides, stachyose, raffinose and yeast cell walls (mannan-oligosachharides (MOSs)) and other fermentable carbohydrates which are not or only minimally digested in the small intestine (Monsan and Paul, 1995; Flickinger et al., 2003; Collins and Gibson, 1999; Ricke, 2015; Roto et al., 2015). In general, the most common oligosaccharides are inulin and its hydrolysates and oligofructans, present in chicory, onion, garlic, asparagus, artichoke, leek, bananas, tomatoes and many other plants (Flickinger et al., 2003; Ricke, 2015). Such prebiotic oligosaccharides can be produced by extraction from plant sources, microbial synthesis or enzymatic hydrolysis of polysaccharides (Crittenden and Playne, 1996; Ricke, 2015). In recent years, use of prebiotics has become a promising approach and offers a potential alternative to growth-promoting antibiotics especially in the monogastric small intestine (Yang et al., 2009; Hume, 2011; Callaway and Ricke, 2012; Alloui et al., 2013; Ricke, 2015; Roto et al., 2015).

2 Beneficial effects of prebiotics: general mechanisms of action

Beneficial effects of prebiotics include altering the GI microbiome, stimulating the immune system, preventing colon cancer, reducing growth and host invasion by such bacteria as *Salmonella* and pathogenic *Escherichia coli* (Cummings and Macfarlane, 2002; Bengmark, 2012; Roto et al., 2015). In general, lactobacilli and bifidobacteria are the two common microbial groups thought to be specifically stimulated by prebiotic supplements (Kaplan and Hutkins, 2000; Ricke, 2015).

Prebiotics can act by physically binding pathogens, increasing the osmotic potential in the lumen of the intestine and/or through metabolites that are produced by the intestinal microbiome which use the prebiotic compounds as specific substrates for their own metabolism (Hajati and Rezaei, 2010). Carbohydrates and proteins indigestible to the host undergo microbiota-dependent hydrolysis and degradation in the lower intestine to produce SCFA and branched chain fatty acids, respectively, that generates Adenosine Triphosphate (ATP) via fermentation pathways (Schell et al., 2002; Ten Bruggencate et al., 2003; Apajalahti, 2005; Rossi et al., 2005; Ricke, 2015; Roto et al., 2015). Even when not utilized for energy production, some prebiotics such as MOSs exert effects to the host by binding to the mannose receptors and inhibit pathogen colonization of the gut (Spring et al., 2000; Patterson and Burkholder, 2003). Overall, there is evidence that incorporating prebiotics as feed supplements facilitates antagonistic effects on enteric pathogens by competitive exclusion, promotion of enzyme reactions (Hajati and Rezaei, 2010), immune stimulation (Monsan and Paul, 1995; Patterson and Burkholder, 2003) and an increase in resistance to pathogen colonization (Bengmark, 2001; Patterson and Burkholder, 2003; Ricke, 2015). Factors influencing the effect of prebiotics include type of diet (amount of non-digestible carbohydrates), type and level of supplements used, animal parameters (species, age, sex and stage of production) (Yang et al., 2009) and hygiene status of the farm (Verdonk et al., 2005; Yang et al., 2009). Specific actions of particular oligosaccharides displaying prebiotic properties are discussed in more detail in the following sections.

3 Non-digestible carbohydrates as prebiotics

These types of carbohydrates are non-digestible in the small intestine; however, they are hydrolysed by colonic bacteria of the large intestine and cecal bacteria depending on the host (Józefiak et al., 2004; Ricke, 2015). The main categories of non-digestible oligosaccharides (NDOs) commonly used in prebiotic supplements include carbohydrates whose monosaccharide units are fructose, galactose, glucose and/or xylose (Bird, 1999).

Indigestible carbohydrates can act as prebiotics when they undergo saccharolytic fermentation in the hindgut and ceca to produce SCFA (Bird, 1999; Józefiak et al., 2004; Apajalahti, 2005; Bird et al., 2010; Hutkins et al., 2016). These fatty acids exhibit beneficial effects including lowering the pH of the environment, improving mineral absorption, inhibiting acid-sensitive pathogens and enhancing the proliferation of indigenous bacteria such as bifidobacteria (Patterson and Burkholder, 2003; Apajalahti, 2005; Bengmark, 2012; Roto et al., 2015).

Non-digestible carbohydrates include NDOs, non-starch polysaccharides and resistant starch (Bird, 1999; Bird et al., 2010; Hutkins et al., 2016). In the beginning, NDOs were used as low-calorie, low cariogenic, sucrose substitutes and bulking agents in the food industry (Crittenden and Playne, 1996; Ricke, 2015). Commercially available prebiotics include FOS, lactulose, isomaltooligosaccharides (IMO), galactooligosaccharides (GOS), TOS, inulin and oligofructose (Collins and Gibson, 1999). Except for FOS, these compounds do not necessarily meet the traditional criteria for prebiotics originally stated by Gibson and Roberfroid (1995) because no specific mechanism of action of colonic fermentation of these complex carbohydrates was known. However, as the gut microbiota and their various metabolic interactions have become better understood, the definition of prebiotics has evolved and now also includes a much wider range of compounds such as NDOs (Bird, 1999; Patterson and Burkholder, 2003; Roberfroid, 2000, 2007; Bird et al., 2010; Bengmark, 2012; Ricke, 2015; Hutkins et al., 2016).

There has been an increasing trend in poultry nutrition research to formulate feeds rich in NDOs that can act as prebiotic supplements (Patterson and Burkholder, 2003; Hajati and Rezaei, 2010; Alloui et al., 2013). Since these relatively resistant oligosaccharides escape the small intestine and can reach the ceca largely intact, they could potentially support the growth of beneficial bacteria including bifidobacteria and lactobacilli as well as other fermentative bacteria and thus exhibit prebiotic properties, as described in the previous section (Bird, 1999; Gomes and Malcata, 1999; Józefiak et al., 2004; Ricke et al., 2013; Lindberg, 2014). As non-traditional cereal grains, cereal coproducts and protein crops have become a greater part of the primary nutritional sources for energy and protein in poultry, increases in the intake of fibre and non-starch carbohydrates may have detrimental impacts on bird performance (Knudsen, 2014); however, with additional research on further processing and extraction approaches, these feedstuffs may offer opportunities to recover components with prebiotic properties. For example, Herfel et al. (2013) demonstrated that stabilized rice bran improved weanling pig performance by exhibiting prebiotic characteristics and Kumar et al. (2012) reported a reduction of S. Typhimurium faecal shedding and an increase in colonization of indigenous lactobacilli in mice fed with rice bran. The opportunity to isolate such components and screen them in vitro under conditions that simulate the GI tract as well as more complete analyses of gut contents (metabolomics, proteomics and others) from poultry fed diets with these dietary ingredients should help to better refine the prebiotic characteristics and develop more targeted benefits (Bird, 1999; Rastall and Maitin, 2002; Donalson et al., 2007, 2008a; Dunkley et al., 2007a,b; Bengmark, 2012; Park et al., 2013; Ricke et al., 2013; Lindberg, 2014).

4 Fructooligosaccharides

FOSs are one of the more frequently used prebiotics (Flickinger et al., 2003; Ricke, 2015). Common sources include onions, Jerusalem artichokes, bamboo shoots, chicory roots and bananas (Flickinger et al., 2003; Ricke, 2015). FOSs are non-digestible and non-absorbable oligosaccharides consisting of short-chain polymers of 1 to 2 linked fructose units (short-chain FOS, SCFO) that can be produced commercially either by hydrolysis of inulin or by enzymatic synthesis from sucrose or lactose (Patterson and Burkholder, 2003; Ricke, 2015). FOSs are resistant to enzymatic degradation and absorption due to the presence of β-linkages, and hence pass through the upper GI tract to reach the colon ceca, where they undergo fermentation (Gibson and Roberfroid, 1995; Gibson, 1999; Flickinger et al., 2003; Xu et al., 2003; Józefiak et al., 2004). The end products of FOS fermentation include SCFAs, which have the ability to inhibit *Salmonella*, *E. coli* and *Clostridium perfringens* (Cummings et al., 2001; Cummings and Macfarlane, 2002; Flickinger et al., 2003; Patterson and Burkholder, 2003; Ricke 2003a). In addition, FOS can act as a fermentable substrate and thus select beneficial intestinal bacteria such as bifidobacteria and lactobacilli (Gibson and Wang, 1994; Ricke, 2015), which, in turn, reduce the pathogens by competitive exclusion (Gibson and Roberfroid, 1995; Flickinger et al., 2003; Xu et al., 2003; Ricke, 2015).

The use of FOS in poultry has been reviewed extensively elsewhere (Ricke, 2015) and will be discussed only briefly in the current review. FOSs can be selectively fermented by most strains of bifidobacteria and support proliferation of bifidobacteria in poultry (Kaplan and Hutkins, 2000; Rossi et al., 2005; Fukuda et al., 2011; Ricke, 2015). Bifidobacteria are anaerobic, gram-positive bacteria found in the GI tract of various warm-blooded animals. The presence of bifidobacteria has been demonstrated to suppress potential pathogens by producing antimicrobials or by lowering the pH through the rapid production of organic acids such as acetate (Fukuda et al., 2011).

Numerous studies have demonstrated the ability of FOS supplementation to reduce the colonization of *Salmonella* in the GI tract (Bailey et al., 1991; Fukata et al., 1999; Donalson et al., 2008b; Ricke, 2015). Dietary amendments to limit *Salmonella* colonization and infection in commercial birds has proven to be particularly critical for laying hens undergoing induced moults (Ricke 2003b; Ricke et al., 2013). Historically, the moulting of commercial laying hens was initiated by complete removal of feed for a number of days to shut down the reproductive tract and cease egg production as a prelude to starting an additional egg laying cycle (Ricke, 2003b). However, emptying of the intestinal tract creates a microenvironment favourable for *S.* Enteritidis colonization and subsequent systemic invasion into the reproductive tract (Durant et al., 1999; Ricke, 2003b; Ricke et al., 2013). Supplying low-energy high-fibre diets such as alfalfa during the moult period proved to be effective in limiting *Salmonella* colonization and infection (Ricke et al., 2013). Although Donalson et al. (2008b) reported that laying hens fed with alfalfa-based diets supplemented with FOS reduced the colonization of *S.* Enteritidis in ovaries and livers when compared to the hens subjected to feed withdrawal, some inconsistencies among independent trials were observed. Clearly, more needs to be understood about the impact of FOS on not only the gut microbiota, but also the physiological status of the gut architecture, along with parameters such as passage rate, and how this impacts the residence time of dietary FOS (Ricke, 2015). For example, Hanning et al. (2012) reported increased villi length and intestinal crypt depth in pasture flock-raised Naked Neck birds fed with FOS, compared to control birds by the end of an eight-week growth cycle.

In a few studies, FOS in their various forms have been used to limit establishment of certain pathogens in poultry. Short-chain FOS and MOS supplementation (4 g/kg) of a dextrose-isolated soy protein diet decreased cecal *C. perfringens* population levels in some of the trials reported by Biggs et al. (2007). When feeding male broilers, Xu et al. (2003) observed that adding 4.0 g/kg FOS enhanced the growth of *Bifidobacterium* and

Lactobacillus, but inhibited *E. coli* populations in small intestinal and cecal contents. Yusrizal and Chen (2003) reported that supplementation of broiler diets with fructans resulted in an increase in lactobacilli counts in the gut and decrease in *Campylobacter* and *Salmonella* in some of the reported trials.

Over time, the mechanisms by which FOS select for beneficial bacteria and create an adverse gut environment for *Salmonella* have become better understood, but the same is not true for other pathogens. In part, this is the case because fewer studies have been done that looked at bacteria other than *Salmonella*. It is important that the effect of FOS on these pathogens also be explored, because the interactions of the indigenous microbiome and these pathogens may differ from that experienced by *Salmonella*. For example, Indikova et al. (2015) have suggested that *Campylobacter* may, to some extent, actually be protected by the microbial consortia that is associated within mixed microbial populations. If these relationships consistently exist in the GI tract, it will be critical to identify the individuals among these potentially supportive microbial communities and use this information to design prebiotics that more effectively target either members antagonistic to *Campylobacter* or that limit those members that are most mutualistically supportive of *Campylobacter*.

5 Yeast-derived components and mannan derivatives as pro- and prebiotics

Yeasts are eukaryotic and unicellular fungi that have been used for both preventive and therapeutic effects against diarrhoea and other GI disturbances (Auclair, 2000; Roto et al., 2015). *Saccharomyces cerevisiae* cells have been extensively examined and commercialized as a potential probiotic source because of their long history in the food and beverage industry that made them readily available (Roto et al., 2015). Live yeast cells gained early popularity as an animal feed additive with various health and animal performance benefits associated with administration either alone or in combination with other probiotic bacteria (Bradley et al., 1994; Auclair, 2000; Fonty and Chaucheyras-Durand, 2006; Roto et al., 2015). Most of the work has been focused on live yeasts as probiotics for ruminants (Roto et al., 2015). Several benefits to rumen function have been noted, including positive impacts on fibre-degrading bacteria, stabilization of rumen pH and enhancement of lactate-utilizing ruminal bacteria (Jouany, 2001; Fonty and Chaucheyras-Durand, 2006).

The presence of lactate-utilizing ruminal bacteria can be critical for ruminants under certain dietary conditions. Lactate accumulation in the rumen occurs after sudden dietary shifts to highly fermentable carbohydrates that can be readily fermented by lactate producers such as *Streptococcus bovis* and other lactic acid bacteria (Herrera et al., 2009). If left unchecked, lactic acid accumulation can become a severe condition in ruminants leading to lactic acidosis, acidification of the rumen and in some cases death of the animal (Herrera et al., 2009). When the lactate-utilizing ruminal bacterium *Selenomonas ruminantium* (Ricke et al., 1996) was co-cultured with a commercial *S. cerevisiae* filter-sterilized filtrate, Nisbet and Martin (1991) observed increases on growth and uptake of radiolabelled lactate by the selenomonad. More recently, Ding et al. (2014) demonstrated that when live *S. cerevisiae* were included in the diets of beef steers, the percentages of lactate-producing bacteria were reduced, while the percentage of *Selenomonas ruminantium* was increased.

In addition to serving as probiotics, yeast cell components can be used as prebiotics. The principal components, glucans and mannans, present in the yeast cell wall along with the fermentation and metabolite products have been extensively examined for prebiotic properties (Baurhoo et al., 2009; Roto et al., 2015). The *S. cerevisiae* cell wall consists of an outer layer composed mostly of glycosylated mannoproteins (90% carbohydrate) and an inner layer (50–60% of the cell wall dry weight) composed of β 1,3-glucan and chitin (Klis et al., 2002). Yeast mannans that are closely associated with cell wall proteins are one of the primary soluble polysaccharides that can be extracted from the cell wall (Lee and Ballou, 1965; Stewart et al., 1968). Although mannan structures are yeast species specific, they are generally characterized as macromolecular polymers in which the mannose backbone is linked via α-(1-6) bonds with side chain-bearing short chains (average of 2–3 mannose units containing α-(1-2) and α-(1-3) linkages) attached to the backbone via α-(1-2) linkages (Jones and Ballou, 1968; Raschke et al., 1973).

MOSs as an isolated component of the yeast outer cell wall were introduced as a feed additive in the early 1990s (Hooge, 2004). The exact mechanism of inhibition of pathogenic bacteria is still uncertain, but several general modes of actions have been proposed and can be summarized as follows: (a) MOS may bind the bacteria with type-1 fimbriae, thereby inhibiting them from binding to carbohydrate moieties of the intestinal lining (Hooge, 2004). Steer et al. (2000) has pointed out that this may be a key mechanism, as most pathogens possess carbohydrate-specific binding proteins for attachment to host cells, and any compound that interferes with that cannot only potentially prevent colonization, but also dislodge already attached cells. (b) Agglutination or clumping of pathogenic cells with type-1 fimbriae and subsequent expulsion from the incubation solution, thus reducing subclinical or lethal infection (Spring et al., 2000; Steer et al., 2000). (c) Immune system modulation (Tizard et al., 1989; Hajati and Rezaei, 2010; Alloui et al., 2013). (d) Modulation of intestinal brush border and mucin enzyme expression (Hooge, 2004; Hajati and Rezaei, 2010). (e) Mannan-derivative induction of phenotypic and functional maturation in mouse dendritic cells (Sheng et al., 2006). The immune modulatory effects of MOS have been attributed to several aspects, as summarized by Tizard et al. (1989), including (1) binding to mannose-binding proteins, (2) attenuating presentation of antigens (from pathogens) to activate macrophages and (3) induction of interleukin-I release.

As a prebiotic, MOS has been examined for both bird health and performance as well as resistance to colonization by food-borne pathogens. Based on an overall statistical meta-analysis across broiler pen trials conducted globally from 1993 to 2003, Hooge (2004) concluded that dietary supplementation of MOS decreased mortalities, while achieving similar body weights and feed conversion ratios (FCRs) when compared to birds receiving antibiotic-supplemented diets. This retention of bird performance appears to hold true when MOS has been combined with immuno-based treatments such as vaccines or introduction of infective microorganisms. For example, Yang et al. (2008b) reported improvements in FCR and/or body weight gains in birds fed with MOS and challenged with pathogenic *E. coli*, compared to the negative control. Likewise, when Nollet et al. (2007) fed MOS to birds vaccinated for *Eimeria* coccidia, followed by a coccidian challenge, they found that MOS improved the FCR of the vaccinated and *Eimeria*-challenged broilers from 15 to 42 days of age, compared to the birds receiving the same coccidia/vaccination combination, but no MOS while remaining the same as the FCR for non-vaccinated/non-challenged birds.

Oyofo et al. (1989a,b) were able to demonstrate that D-mannose could block *in vitro* attachment on chicken small intestinal epithelial cells and reduce cecal colonization of

S. Typhimurium in challenged birds. However, MOSs impact on cecal microbiota and the cecal microenvironment has been somewhat inconsistent. Spring et al. (2000) did see some reductions in S. Typhimurium and S. Dublin-challenged broiler chicks fed with MOS, but no effect on cecal concentrations of lactobacilli, enterococci, anaerobic bacteria, lactate, SCFA or cecal pH. Populations of lactobacilli and coliforms were reduced in the ileum of birds fed with 2 g/kg of MOS, but less impact was observed in the ceca (Yang et al., 2008a). Environmental management conditions may be a factor as well. For example, when birds were raised in the presence of new litter and fed with 2 g/kg of MOS, Yang et al. (2007) did not detect any differences in total anaerobes, C. perfringens or lactic acid bacterial populations in the ileum, duodenum or ceca. Using a C. perfringens challenge model, Thanissery et al. (2010) observed only a slight reduction of intestinal C. perfringens in broiler chicks fed with a commercial yeast extract on days 1 and 7 post-challenge.

6 Galactooligosaccharide and isomaltooligosaccharide

GOSs are generated from lactose by a glycosyl transfer of one or more D-galactosyl units onto a chain of galactose units (Crittenden and Playne, 1996; Mahoney, 1998; Boon et al., 2000; Tzortzis et al., 2005; Macfarlane et al., 2008). The composition of GOS varies depending on the degree of polymerization ranging from two to eight monomeric units (Macfarlane et al., 2008). The impact of GOS on the host may depend on a number of factors, but they are believed to preferentially stimulate the growth of organisms such as bifidobacteria and lactobacilli by providing a fermentable substrate and thus generating an inhibitory GI microenvironment for pathogens (Gomes and Malcata, 1999; Rastall and Maitin, 2002; Tzortzis et al., 2005; Macfarlane et al., 2008; Saminathan et al., 2011; Alloui et al., 2013). Other inhibitory mechanisms may be in play as well. Shoaf et al. (2006) demonstrated that GOS could reduce adherence of enteropathogenic E. coli to epithelial cell tissue cultures, leading them to suggest that GOS could act as a structural mimic of the pathogen binding sites on the GI epithelial surface. However, such binding may be both host- and pathogen strain-specific as Naughton et al. (2001) did not observe decreases in S. Typhimurium or non-pathogenic E. coli association with pig intestinal cells when a commercial GOS product was included. Intestinal architecture could also be influenced, as Hanning et al. (2012) reported that at eight weeks pasture, flock-raised Naked Neck broilers fed with GOS exhibited the deepest intestinal crypt depth of the intestinal villi, when compared to birds fed either with FOS, plum fibres or with no feed additive.

IMOs are glucosyl saccharides linked by α 1-6 glucosidic bonds and occur in the form of compounds such as isomaltose, panose and isomaltotriose (Kuriki et al., 1993). They possess a degree of polymerization anywhere from 2 to 6 and can be derived from a variety of sources, including starch, maltose, dextran and sucrose (Crittenden and Playne, 1996; Aslan and Tanriseven, 2007). IMOs have been proposed as a potential prebiotic for poultry use (Chung and Day, 2004). Based on this premise, Chung and Day (2004) screened a branched α-IMO produced by Leuconostoc mesenteroides and demonstrated that unidentified poultry cecal isolates as well as Bifidobacterium longum could use the IMO but E. coli and S. Typhimurium could not. When cecal isolates and S. Typhimurium were grown in mixed culture with IMO, the Salmonella populations were reduced (Chung and Day, 2004). In a follow-up in vivo study, broilers challenged with S. Typhimurium and fed either 1, 2 or 4% wt/wt IMO exhibited a 2-log CFU/ml reduction of cecal Salmonella only at the 1% IMO level versus the control (no IMO), while an increase in cecal bifidobacteria populations was similar for all levels of IMO (Thitaram et al., 2005). However, Zhang et al. (2003) reported that while IMO inclusion in broiler diets improved performance during the first three weeks, no detectable changes occurred in the levels of *Lactobacillus*, *E. coli* and total aerobes from the crop and the ceca.

7 Guar gum as a potential prebiotic source

Guar gum is a highly viscous soluble fibre produced from the seed of the Indian cluster bean (*Cyamopsis tetragonoloba*) and has been used widely in the food industry as an emulsion stabilizer and thickener (Tuohy et al., 2001; Slavin and Greenberg, 2003). Guar gum is a polysaccharide, consisting of a 1-4 linked β-D-mannopyranose backbone with a branched 1-6-α-D-galactopyranose. Guar meal is a by-product of guar gum manufacture and exhibits two deleterious properties, namely, it acts as a trypsin inhibitor and contains residual gum (approximately 18–20%) (Couch et al., 1966; Nagpal et al., 1971; Lee et al., 2003). Since guar gum is a highly branched polysaccharide, it can be categorized in the 'well-fermentable' fibre category which includes pectin, acacia, polydextrose, inulin and oligosaccharides, which when fermented, result in large quantities of SCFAs, including acetic, butyric and propionic acids in varying proportions (Bengmark, 1998; Bird, 1999; Józefiak et al., 2004; Ricke et al., 2013; Lindberg, 2014; Hutkins et al., 2016).

Feeding guar gum and guar meal to broilers has reportedly caused growth depression (Vohra and Kratzer, 1964a,b; Verma and MacNab, 1982). Over the years, this was explained by the presence of gum residue as the former is responsible for delayed gastric emptying and increased intestinal transit time because of its naturally high viscosity (Anderson and Warnick, 1964; Vohra and Kratzer, 1964a,b; Blackburn and Johnson, 1981). Guar gums also elicit some beneficial effects in some hosts such as decreased plasma cholesterol (Moriceau et al., 2000; Yamamoto et al., 2000), decreased postprandial serum glucose (Fairchild et al., 1996; Ou et al., 2001) and suppression of colon cancer (Heitman et al., 1992).

Guar meal is a processed by-product of guar beans rich in crude protein and consists of two fractions: a high protein germ fraction and a low protein psyllium husk (Van Etten et al., 1961; Couch et al., 1967; Lee et al., 2003). The gum content of typical guar meals is anywhere from 18 to 20% and contains the trypsin inhibitor (Couch et al., 1966; Nagpal et al., 1971). When considered for meeting amino acid requirements of poultry, analysis of guar meal reveals that it is rich in arginine, but low in sulphur amino acids (Verma and MacNab, 1984). Leucine and isoleucine are low when compared to soya bean and groundnut protein sources (Verma and MacNab, 1984). By and large, the germ fraction is considered the more beneficial fraction compared to the hull fraction for inclusion in broiler diets (Lee et al. 2003). Feeding high concentrations (10–15%) of guar meal caused diarrhoea, depressed growth and increased mortality in broilers, and reduced egg production in layers (Patel and McGinnis, 1985), but was an effective moult induction diet for initiating a second egg production cycle in laying hens (Zimmermann et al., 1987). The upper limit for feeding guar germ fraction to broiler chickens can be increased somewhat with the addition of an industrial source of β-mannanase in the feed (Lee et al., 2005).

Partially hydrolyzed guar gum (PHGG), a mixture of galactomannans, low in molecular weight and viscosity, is obtained by selectively cleaving the backbone chain of the guar gum using endo-β-D-mannanase and can be added to feed formulations without retaining the deleterious effects of the gum residue (Slavin and Greenberg, 2003; Yoon et al., 2008). PHGG has demonstrated therapeutic effects including reducing incidence or duration of diarrhoea (Hohmann et al., 1994; Slavin and Greenberg, 2003; Alam et al., 2000, 2005) and improving the GI microbiome by selection of beneficial microorganisms such as bifidobacteria and lactobacilli spp. in human subjects (Okubo et al., 1994; Tuohy et al., 2001). Feeding PHHG (0.025%) to pullets for six weeks proved to be beneficial as it decreased S. Enteritidis colonization in internal organs and intestine, reduced incidence on the egg surface and increased the cecal bifidobacteria and lactobacilli spp. (Ishihara et al., 2000). Moreover, feeding Sprague–Dawley rats with PHHG enhanced IgA productivity in the spleen and mesenteric lymph node lymphocytes, when compared to cellulose-fed rats (Yamada et al., 1999).

8 Synbiotics: combining pre- and probiotics for enhanced nutritional supplements

As there is evidence that probiotics can have similar effects on bird performance, health and prevent colonization by food-borne pathogens as prebiotics, it makes sense to combine selected pre- and probiotics to achieve an even more effective feed additive. Since many prebiotics serve as substrates for certain GI organisms, their effects might be enhanced by combining them with selective probiotic cultures that would employ synergism in their action for better protective effects against food-borne pathogens. These combinations have been referred to as synbiotics and are defined as nutritional supplements which contain a mixture of prebiotics and probiotics that act synergistically and deliver beneficial effects to the host by improving the survival and establishment of live microbial dietary supplements in the GI tract (Gibson and Roberfroid, 1995; Schrezenmeir and de Vrese, 2001; Rastall and Maitin, 2002; Patterson and Burkholder, 2003; Rastall, 2004; Bengmark, 2012). However, achieving synergistic effects in the host requires developing optimal combinations of prebiotic and probiotic (Rastall et al., 2005).

While conceptionally the idea of a symbiotic appears straightforward, mechanistically it can be more complicated. This is because inclusion of a prebiotic involves not only manipulation of the indigenous microbiome and host responses such as the immune system, but the type of prebiotic that is provided as well as the particular bacteria serving as the target probiotic (Bomba et al., 2002; Bengmark, 2012). For example, *in vitro* mixed culture studies conducted by Fooks and Gibson (2002) demonstrated that oligosaccharides supported growth of *Lactobacillus plantarum* and *Bifidobacterium bifidum* and inhibited *E. coli*, *Campylobacter jejuni* and S. Enteritidis, but the extent of reduction was dependent on the type of oligosaccharide. Herrera et al. (2012) not only noted a similar carbohydrate specificity in *in vitro* co-cultures of *Streptococcus bovis* and S. Typhimurium, but also observed that allowing time for the *Streptococcus bovis* to grow and ferment the carbohydrate prior to inoculating S. Typhimurium was critical to maximizing reduction of the pathogen. Based on their *in vitro* screening study with poultry *Lactobacillus* isolates, Saminathan et al. (2011) concluded that not only were there differences in isolate responses to different prebiotic substrates, but there were strain differences in the ability to grow on a particular prebiotic.

A novel way to optimize prebiotic/probiotic combinations is to simply employ the prebiotic to selectively isolate GI tract bacteria that then become probiotic candidates possessing the desired properties of interest in the corresponding host animal. A classic example of this approach is using lactose as the carbon source to isolate a defined cecal population that would limit *Salmonella* colonization in poultry flocks. Lactose as a poultry feed candidate to target GI tract bacteria is attractive because its utilization by the bird is generally low at all times and it cannot be used by *S.* Typhimurium (Atkinson et al., 1957; Gutnick et al., 1969; Moran, Jr., 1985; Waldroup et al., 1992; Kermanshahi and Rostami, 2006). Lactose either directly fed, added to water or as a part of a feed supplement such as whey, was initially shown to inhibit *Salmonella* colonization in broilers and believed to be fermented by cecal bacteria to generate organic acids inhibitory to *Salmonella* (Bilgili and Moran, Jr., 1990; Corrier et al., 1990a,b; DeLoach et al., 1990; Kermanshahi and Rostami, 2006). The relationship between the cecal microbiome and lactose was further substantiated when dietary lactose was combined with anaerobic cecal bacterial inocula and introduced to broiler chickens and turkey poults (Hinton, Jr. et al., 1991; Ziprin et al., 1990; Corrier et al., 1991a,b). The cecal inocula had been generated by incubating a mixed cecal culture for 24 h in a Viande–Levure (VL) broth with lactose and no glucose (Corrier et al., 1991b). In order to stabilize the cecal inocula, Nisbet et al. (1993a,b) took this a step further and used continuous flow (CF) cultures containing lactose VL broth in place of glucose to select for a cecal lactose-utilizing consortium that stabilized after reaching steady state following several growth vessel turnovers. When they inoculated chickens, significant *S.* Typhimurium reduction was achieved with the CF culture combined with lactose, when compared to control birds, and these birds also yielded higher levels of cecal propionic acid (Nisbet et al., 1993b). As further refinements were introduced to avoid dietary dependence on relatively expensive lactose, it eventually became clear that developing and maintaining a metabolically stable cecal microbial consortium that could consistently generate propionic acid in the ceca of inoculated birds was a critical element for achieving potential commercial success (Corrier et al., 1995; Nisbet et al., 1996a,b).

Other symbiotic combinations have also proven to be beneficial to the poultry host and/or inhibitory to pathogen colonization and some have since been commercialized. These studies are too numerous to list and discuss here. However, overall outcomes tend to vary with the type of application and composition of the symbiotic. This is not necessarily surprising, given the extra layers of complexity associated not just with the different chemical compositions of individual prebiotics but also considering the biological variability associated with the probiotic live cultures, and finally the immense and still poorly understood GI tract microbiome. Host factors such as the immune response to a shifting GI tract microbiome as the host animal matures are critical as well. The interaction between the prebiotic and the probiotic introduces additional mechanistic complexities related to stability of the prebiotic in the digestive tract and its relative fermentability by the GI microbiome. This interaction has potential commercial implications. One of the benefits of including prebiotics as part of the feed amendment is the potential to protect probiotic bacteria from hostile host environments such as the acidic gastric compartment as they travel through the upper intestinal tract to the target areas of the lower GI tract (Rastall and Maitin, 2002). Obviously, the degree of protection will potentially influence the consistency of the symbiotic.

Finally, the response of individual pathogen strains to external environmental stresses and interventions employed in the food processing industry are not always the same and strain variation must be taken into account (Lianou and Koutsoumanis, 2013). This no doubt holds true for *Salmonella* responses to the presence of prebiotics and synbiotics as well. For example, *Salmonella* serovar and strain differences in susceptibility to antagonistic metabolic products such as fermentation acids generated by probiotic cultures do occur and therefore should probably be taken into account (Ricke, 2003a; Dunkley et al., 2009; Foley et al., 2013; Ricke, 2014). As interest in prebiotics and synbiotics grows, it will be important to screen the candidates against not only multiple *Salmonella* serovars but several representative strains for each individual serovar as well.

9 Summary and conclusions

Prebiotics are a class of feed ingredients that can act as substrates in the host to produce various health benefits. Obviously, as more has become known about the interactions and activities occurring in the host microbiome, this definition has evolved considerably and will continue to do so. Regardless of the definition, the goal of prebiotic supplementation remains, namely improving the growth performance, health status and decreasing the food-borne pathogen load of the host animal. As antibiotics are removed from the diets of food animals, achieving these goals with non-antibiotic alternatives will become commercially imperative. Likewise with the increasing interest in non-traditional animal production such as organic and natural animal husbandry, such alternatives will be particularly attractive to the consumer of these products (Ricke et al., 2012b). Optimizing combinations of prebiotics and probiotic cultures into synbiotics offers an opportunity to enhance the growth of beneficial bacteria such as bifidobacteria and lactobacilli spp. in a more controlled fashion that will ultimately produce more predictable synergistic effects in the host. However, such combinations, while conceptually quite attractive, are fraught with complexities that require a better understanding not only of the GI microbiome, but of the host animal responses as well. Likewise, further research is needed to gain a better understanding of the mechanism(s) of action against food-borne pathogens such as *Salmonella* in the GI microbiome, and the pathogen's capabilities to express resistance, and variability among strains within a pathogen species.

Other factors need to be optimized as well to achieve more effectiveness for these types of products and to make them more commercially attractive. Certainly, further chemical refinement of prebiotic compounds to be more mechanistically selective and consistently predictable towards targeted GI tract microorganisms would be helpful. For example, Maillard reaction products may offer some interesting multifunctional properties including antimicrobial activities directed against food-borne pathogens (Ricke, 2015). Chalova et al. (2012) used transcriptomic microarrays to demonstrate that *S.* Typhimurium growing on lysine–glucose Maillard compounds expressed genes related to an energy starvation mode. This is an outcome that offers intriguing potential, and as genomic information further expands for both pathogens and GI bacteria the ability to refine such compounds to favour the beneficial bacteria while simultaneously being antagonistic to the pathogen population could be explored and commercially developed. Likewise, designing prebiotics that are more resistant to degradation until they reach their respective target GI tract compartment would probably improve efficacy. These chemical modifications would in part support an increasing interest in more novel sources of NDO-based prebiotics.

Probably one of the areas of most interest is retaining efficacy of the biologically complex product during all phases of its production and eventual delivery to the target group of food animals. Using advanced technologies such as biofunctionalized carriers or nanoparticles (Taylor et al., 2004; Luo et al., 2005) that could serve as multifunctional/multipurpose entities and thus be used to retain diverse activities ranging from prebiotic properties to multiple hurdle antimicrobial activities (Ricke and Hanning, 2013) that exclusively target food-borne pathogens is attractive. Multifunctional carriers offer a conceptually attractive 'next generation' approach to further optimize the efficacy and precision for prebiotics. However, cost-effectiveness for generating such particles on a large scale, as well as regulatory issues, will have to be considered before these can be practical for food animal production systems. Such particles would probably also need to be environmentally friendly, which means they would need to be biodegradable with a relatively short half-life. Regardless of the potential future research directions, prospects appear to be promising for prebiotics playing a major commercial role in poultry and animal production systems.

10 Where to look for further information

Callaway, T. R. and Ricke, S. C. (eds) (2012), *Direct Fed Microbials/Prebiotics for Animals: Science and Mechanisms of Action*. Springer Science, New York, NY.

Gibson G. R. and Roberfroid, M. B. (1995), Dietary modulation of the human colonic microflora: Introducing the concept of prebiotics. *J. Nutr. 125*, 1401–12.

Hutkins, R. W., Krumbeck, J. A., Bindels, L. B., Cani, P. D., Fahey, Jr., G. C., Goh, Y. J., Hamaker, B., Martens, E. C., Mills, D. A., Rastal, R. A., Vaughan, E. and Sanders, M. E. (2016), Prebiotics: Why definitions matter. *Curr. Opin. Biotech. 37*, 1–7.

Patterson, J. A. and Burkholder, K. (2003), Application of prebiotics and probiotics in poultry production. *Poult. Sci. 82*, 627–31.

Rastall, R. (2004), Bacteria in the gut: Friends and foes and how to alter the balance. *J. Nutr. 134*, 2022S–6S.

Rastall, R. A. and Maitin, V. (2002), Prebiotics and synbiotics: Towards the next generation. *Curr. Opin. Biotechnol. 13*, 490–6.

Rastall, R. A., Gibson, G. R., Gill, H. S., Guarner, F., Klaenhammer, T. R., Pot, B., Reid, G., Rowland, I. R. and Sanders, M. E. (2005), Modulation of the microbial ecology of the human colon by probiotics and synbiotics to enhance human health: An overview of enabling science and potential applications. *FEMS Microbiol. Ecology 52*, 145–52.

Ricke, S. C. (2015), Potential of fructooligosaccharide prebiotics in alternative and nonconventional poultry production systems. *Poult. Sci. 94*, 1411–18.

Roberfroid, M. B. (2007), Prebiotics: The concept revisited. *J. Nutr. 137*, 830S–7S.

Roto, S. M., Rubinelli, P. M. and Ricke, S. C. (2015), An introduction to the avian gut microbiota and the effects of yeast-based prebiotic compounds as potential feed additives. *Front. Vet. Sci. 2 (Article 28)*, 1–18. doi: 10.3389/fvets.2015.00028.

Wang, Y. (2009), Prebiotics: Present and future science and technology. *Food Res. Int. 42*, 8–12.

11 References

Alam, N. H., Meier, R., Schneider, H., Sarker, S. A., Bardhan, P. K., Mahalananbis, D., Fuchs, G. J. and Gyr, N. (2000), Partialy hydrolyzed guar gum-supplemented oral rehydration solution in the treatment of acute diarrhea in children. *J. Ped. Gastroenterol. Nutr. 31*, 503–7.

Alam, N. H., Meier, R., Sarker, S. A., Bardhan, P. K., Schneider, H. and Gyr, H. (2005), Partially hydrolysed guar gum supplemented comminuted chicken diet in persistent diarrhoea: A randomised controlled trial. *Arch. Dis. Child. 90*, 195–9.

Alloui, M. N., Szczurek, W. and Świątkiewicz, S. (2013), The usefulness of prebiotics and probiotics in modern poultry nutrition: A review. *Ann. Anim. Sci. 13*, 17–32.

Anderson, J. O. and Warnick, R. E. (1964), Value of enzyme supplements in rations containing certain legume seed meals or gums. *Poult. Sci. 43*, 1091–7.

Apajalahti, J. (2005), Comparative gut microflora, metabolic challenges, and potential opportunities. *J. Appl. Poult. Res. 14*, 444–53.

Aslan, Y. and Tanriseven, A. (2007), Immobilization of *Penicillium lilacinum* dextranase to produce isomaltooligosaccharides from dextran. *Biochem. Eng. J. 34*, 8–12.

Atkinson, R. L., Kratzer, F. H. and Stewart, G. F. (1957), Lactose in animal and human feeding. *J. Dairy Sci. 50*, 1114–32.

Auclair, E. (2000), Yeast as an example of the mode of action of probiotics in monogastric and ruminant species. III Conference show feed manufacturing in the Mediterranean region improving food safety. From feed to food. *Reus, Espana. 54*, 45–53.

Bailey, J. S., Blankenship, L. C. and Cox, N. A. (1991), Effect of fructooligosaccharide on *Salmonella* colonization of the chicken intestine. *Poult. Sci. 70*, 2433–8.

Baurhoo, B., Goldflus, F. and Zhao, X. (2009), Purified cell wall of *Saccharomyces cerevisiae* increases protection against intestinal pathogens in broiler chickens. *Int. J. Poult. Sci. 8*, 133–7.

Bengmark, S. (1998), Immunonutrition: Role of biosurfactants, fiber, and probiotic bacteria. *Nutrition 14*, 585–94.

Bengmark S. (2001), Pre-, pro- and synbiotics. *Curr. Opin. Clin. Nutr. Metab. Care. 4*, 571–9.

Bengmark, S. (2012), Integrative medicine and medicine-The role of pre-, pro- and synbiotics. *Clin. Translational Med. 1:6*, 1–13 (http://www.clintransmed.com/content/1/1/6).

Biggs, P., Parsons, C. M. and Fahey, Jr., G. C. (2007), The effects of several oligosaccharides on growth performance, nutrient digestibilities, and cecal microbial populations in young chicks. *Poult. Sci. 86*, 2327–36.

Bilgili, S. F. and Moran, Jr., E. T. (1990), Influence of whey and probiotic-supplemented withdrawal feed on the retention of *Salmonella* intubated into market age broilers. *Poult. Sci. 69*, 1670–4.

Bird, A. R. (1999), Prebiotics: A role for dietary fibre and resistant starch? *Asia Pacific J. Clin. Nutr. 8 (Suppl.)*, S32–6.

Bird, A. R., Conlon, M. A., Christopher, C. T. and Topping, D. L. (2010), Resistant starch, large bowel fermentation and a broader perspective of prebiotics and probiotics. *Benef. Microbes 1*, 423–31.

Blackburn, N. A. and Johnson, J. T. (1981), The effect of guar gum on the viscosity of the gastrointestinal contents and on glucose uptake from the perfused jejunum in the rat. *Br. J. Nutr. 46*, 239–46.

Bomba, A., Nemcová, R., Gancarčiková, S., Herich, R., Guba, P. and Mudroňová, D. (2002), Improvement of the probiotic effect of micro-organisms by their combination with maltodextrins, fructooligosaccharides and polyunsaturated fatty acids. *Br. J. Nutr. 88 (Suppl. 1)*, S95–9.

Boon, M. A., van't Riet, K. and Janssen, A. E. M. (2000), Enzymatic synthesis of oligosaccharides: Product removal during a kinetically controlled reaction. *Biotechnol. Bioeng. 70*, 411–20.

Bradley, G. L., Savage, T. F. and Timm, K. I. (1994), The effects of supplementing diets with *Saccharomyces cerevisiae* var. *boulardii* on male poult performance and ileal morphology. *Poult. Sci. 73*, 1766–70.

Callaway, T. R. and Ricke, S. C. (eds) (2012), *Direct Fed Microbials/Prebiotics for Animals: Science and Mechanisms of Action*. Springer Science, New York, NY.

Chalova, V. I., Hernández-Hernández, O., Muthaiyan, A., Sirsat, S. A., Natesan, S., Sanz, M. L., Javier Moreno, F., O'Bryan, C. A., Crandall, P. G. and Ricke, S. C. (2012), Growth and transcriptional response of *Salmonella* Typhimurium LT2 to glucose-lysine-based Maillard reaction products generated under low water activity conditions. *Food Res. Int. 45*, 1044–53.

Chen, C.-Y., Yan, X. and Jackson, C. (eds) (2015), *Antimicrobial Resistance and Food Safety: Methods and Techniques*. Academic Press/Elsevier Science, Amsterdam, The Netherlands.

Chung, C.-H. and Day, D. F. (2004), Efficacy of *Leuconostoc mesenteroides* (ATCC 13146) isomaltooligosaccharides as a poultry prebiotic. *Poult. Sci. 83*, 1302–6.

Collins, M. D. and Gibson, G. R. (1999), Probiotics, prebiotics, and synbiotics: Approaches for modulating the microbial ecology of the gut. *Am. J. Clin. Nutr. 69 (Suppl),* 1052S–7S.

Corrier, D. E., Hinton, Jr., A., Ziprin, R. L., Beier, R. C. and DeLoach, J. (1990a), Effect of dietary lactose on cecal pH, bacteriostatic volatile fatty acids, and Salmonella typhimurium colonization of broiler chicks. *Avian Dis. 34,* 668–76.

Corrier, D. E., Hinton, Jr., A, Ziprin, R. L. and DeLoach, J. (1990b), Effect of dietary lactose on Salmonella colonization of market-age broiler chickens. *Avian Dis. 34,* 668–76.

Corrier, D. E., Hargis, B., Hinton, Jr., A., Lindsey, D., Caldwell, D., Manning, J. and DeLoach, J. (1991a), Effect of anaerobic cecal microflora and dietary lactose on colonization resistance of layer chicks to invasive Salmonella enteritidis. *Avian Dis. 35,* 337–43.

Corrier, D. E., Hinton, Jr., A., Kubena, L. F., Ziprin, R. L. and DeLoach, J. (1991b), Decreased *Salmonella* colonization in turkey poults inoculated with anaerobic cecal microflora and provided dietary lactose. *Poult. Sci. 70,* 1345–50.

Corrier, D. E., Nisbet, D. J., Scanlan, C. M., Hollister, A. G. and DeLoach, J. R. (1995), Control of *Salmonella typhimurium* colonization in broiler chicks with a continuous-flow characterized mixed culture of cecal bacteria. *Poult. Sci. 74,* 916–24.

Couch, J. R., Creger, C. R. and Bakshi, Y. K. (1966), Trypsin inhibitor in guar meal. *Proc. Soc. Exp. Biol. Med. 123,* 263–5.

Couch, J. R., Bakshi, Y. K., Ferguson, T. M., Smith, E. B. and Creger, C. R. (1967), The effect of processing on the nutritional value of guar meal for broiler chicks. *Br. Poult. Sci. 8,* 243–50.

Crittenden, R. G. and Playne, M. J. (1996), Production, properties, and applications of food-grade oligosaccharides. *Trends Food Sci. Tech. 7,* 353–61.

Cummings, J. H. and Macfarlane, G. T. (2002), Gastrointestinal effects of prebiotics. *Br. J. Nutr. 87,* S145–51.

Cummings, J. H., Macfarlane, G. T. and Englyst, H. N. (2001), Prebiotic digestion and fermentation. *Am. J. Clin. Nutr. 73,* S415–20.

DeLoach, J. R., Oyofo, B. A., Corrier, D. E., Kubena, L. F., Ziprin, R. L. and Norman, J. O. (1990), Reduction of Salmonella typhimurium concentration in broiler chickens by milk or whey. *Avian Dis. 34,* 389–92.

Demigné, C., Yacoub, C. and Rémésy, C. (1986), Effects of absorption of large amounts of volatile fatty acids on rat liver metabolism. *J. Nutr. 116,* 77–86.

Ding, G., Chang, Y., Zhao, L., Zhou, Z., Ren, L. and Meng, Q. (2014), Effect of *Saccharomyces cerevisiae* on alfalfa nutrient degradation characterstics and rumen microbial populations of steers fed diets with different concentrate – to – forage ratios. *J. Anim. Sci. Biotechnol. 5,* 24 (http://www.jasbsci.com/content/5/1/24).

Donalson, L. M., Kim, W. K., Chalova, V. I., Herrera, P., Woodward, C. L., McReynolds, J. L., Kubena, L. F., Nisbet, D. J. and Ricke, S. C. (2007), In vitro anaerobic incubation of *Salmonella enterica* serotype Typhimurum and laying hen cecal bacteria in poultry feed substrates and a fructooligosacharide prebiotic. *Anaerobe 13,* 208–14.

Donalson, L. M., Kim, W. K., Chalova, V. I., Herrera, P., McReynolds, J. L., Gotcheva, V. G., Vidanović, D., Woodward, C. L., Kubena, L. F., Nisbet, D. J. and Ricke, S. C. (2008a). In vitro fermentation response of laying hen cecal bacteria to combinations of fructooligosaccharide (FOS) prebiotic with alfalfa or a layer ration. *Poult. Sci. 87,* 1263–75.

Donalson, L. M., McReynolds, J. L., Kim, W. K., Chalova, V. I., Woodward, C. L., Kubena, L. F., Nisbet, D. J. and Ricke, S. C. (2008b), The influence of a fructooligosaccharide prebiotic combined with alfalfa molt diets on the gastrointestinal tract fermentation, *Salmonella* Enteritidis infection and intestinal shedding in laying hens. *Poult. Sci. 87,* 1253–62.

Dunkley, K. D., Dunkley, C. S., Njongmeta, N. L., Callaway, T. R., Hume, M. E., Kubena, L. F., Nisbet, D. J. and Ricke, S. C. (2007a), Comparison of in vitro fermentation and molecular microbial profiles of high-fiber feed substrates (HFFS) incubated with chicken cecal inocula. *Poult. Sci. 86,* 801–10.

Dunkley, K. D., McReynolds, J. L., Hume, M. E., Dunkley, C. S., Callaway, T. R., Kubena, L. F., Nisbet, D. J. and Ricke, S. C. (2007b), Molting in *Salmonella* Enteritidis-challenged laying hens fed alfalfa crumbles.II. Fermentation and microbial ecology response. *Poult. Sci. 86*, 2101–9.

Dunkley, K. D., Callaway, T. R., Chalova, V. I., McReynolds, J. L., Hume, M. E., Dunkley, C. S., Kubena, L. F., Nisbet, D. J. and Ricke, S. C. (2009), Foodborne *Salmonella* ecology in the avian gastrointestinal tract. *Anaerobe 15*, 26–35.

Durant, J. A., Corrier, D. E., Byrd, J. A., Stanker, L. H. and Ricke, S. C. (1999), Feed deprivation affects crop environment and modulates *Salmonella enteritidis* colonization and invasion of leghorn hens. *Appl. Environ. Microbiol. 65*, 1919–23.

Fairchild, R. M., Ellis, P. R., Byrne, A. J., Luzio, S. D. and Mir, M. A. (1996), A new breakfast cereal containing guar gum reduces postprandial plasma glucose and insulin concentrations in normal-weight human subjects. *Br. J. Nutr. 76*, 63–73.

Finstad, S., O'Bryan, C. A., Marcy, J. A., Crandall, P. G. and Ricke, S. C. (2012), *Salmonella* and broiler production in the United States: Relationship to foodborne salmonellosis. *Food Res. Int. 45*, 789–94.

Flickinger, E. A., Van Loo, J. and Fahey, Jr., G. C. (2003), Nutritional responses to the presence of inulin and oligofructose in the diets of domesticated animals: A review. *Crit. Revs. Food Sci. Nutr. 43*, 19–60.

Foley, S. L., Johnson, T. J., Ricke, S. C., Nayak, R. and Danzeisen, J. (2013), *Salmonella* pathogenicity and host adaptation in chicken-associated serovars. *Microbiol. Mol. Biol. Revs. 77*, 582–607.

Foley, S., Nayak, R., Hanning, I. B., Johnson, T. L., Han, J. and Ricke, S. C. (2011), Population dynamics of *Salmonella enterica* serotypes in commercial egg and poultry production. *Appl. Environ. Microbiol. 77*, 4273–9.

Fonty, G. and Chaucheyras-Durand, F. (2006), Effects and modes of action of live yeasts in the rumen. *Biologia, Bratislava 61/6*, 741–50.

Fooks, L. J. and Gibson, G. R. (2002), In vitro investigations of the effect of probiotics and prebiotics on selected human intestinal pathogens. *FEMS Microbiol. Ecol. 39*, 67–75.

Fukata, T., Sasai, K., Miyamoto, T. and Baba, E. (1999), Inhibitory effects of competitive exclusion and fructooligosaccharide, singly and in combination, on *Salmonella* colonization of chicks. *J. Food Prot. 62*, 229–33.

Fukuda, S., Toh, H., Hase, K., Oshima, K., Nakanishi, Y., Yoshimura, K., Tobe, T., Clarke, J. M., Topping, D. L., Suzuki, T., Taylor, T. D., Itoh, K, Kikuchi, J., Morita, H., Hattori, M. and Ohno, H. (2011), Bifidobacteria can protect from enteropathogenic infection through production of acetate. *Nature 469*, 543–7.

Fuller, R. (1989), Probiotics in man and animals. *J. Appl. Bacteriol. 66*, 365–78.

Gibson, G. R. and Wang, X. (1994), Inhibitory effects of bifidobacteria on other colonic bacteria. *J. Appl. Bacteriol. 77*, 412–20.

Gibson G. R. and Roberfroid, M. B. (1995), Dietary modulation of the human colonic microflora: Introducing the concept of prebiotics. *J. Nutr. 125*, 1401–12.

Gibson, G. R. (1999), Dietary modulation of the human gut microflora using the prebiotics oligofructose and inulin. *J. Nutr. 129*, 1438–41.

Gomes, A. M. P. and Malcata, F. X. (1999), *Bifidobacterium* spp. and *Lactobacillus acidophilus*: Biological, biochemical, technological and therapeutic properties relevant for use as probiotics. *Trends Food Sci. Technol. 10*, 139–57.

Gutnick, D., Calvo, J. M., Klopowski, T. and Ames, B. N. (1969), Compounds which serve as the sole source of carbon or nitrogen for *Salmonella typhimurium* LT-2. *J. Bacteriol. 100*, 215–19.

Hajati, H. and Rezaei, M. (2010), The application of prebiotics in poultry production. *Int. J. Poult. Sci. 9*, 298–304.

Hanning, I., Clement, A., Owens, C., Park, S. H., Pendleton, S., Scott, E. E., Almeida, G., Gonzalez Gil, F. and Ricke, S. C. (2012), Assessment of production performance in two breeds of broilers fed prebiotics as feed additives. *Poult. Sci. 91*, 3295–9.

Heitman, D. W., Hardman, W. E. and Cameron, L. L. (1992), Dietary supplementation with pectin and guar gum on 1,2-dimethylhydrazine-induced colon carcinogenesis in rats. *Carcinogenesis 13*, 815–18.

Herfel, T., Jacobi, S., Lin, X., van Heugten, E., Fellner, V. and Odle, J. (2013), Stabilized rice bran improves weaning pig performance via a prebiotic mechanism. *J. Anim. Sci. 91*, 907–13.

Herrera, P., Kwon, Y. M., Maciorowski, K. G. and Ricke, S. C. (2009), Ecology and pathogenicity of gastrointestinal *Streptococcus bovis*. *Anaerobe 15*, 44–54.

Herrera, P., O'Bryan, C. A., Crandall, P. G. and Ricke, S. C. (2012), Growth response of *Salmonella enterica* Typhimurium in co-culture with ruminal bacterium *Streptococcus bovis* is influenced by time of inoculation and carbohydrate substrate. *Food Res. Int. 45*, 1054–7.

Hinton, Jr., A., Corrier, D. E., Ziprin, R. L., Spates, J. O. and DeLoach, J. R. (1991), Comparison of the efficacy of cultures of cecal anaerobes as inocula to reduce *Salmonella typhimurium* colonization in chicks with or without dietary lactose. *Poult. Sci. 70*, 67–73.

Hohmann, H.-H., Kemen, M., Fuessenenich, C., Senkal, M. and Zumtobel, V. (1994), Reduction in diarrhea incidence by soluble fiber in patients receiving total or supplemental enternal nutrition. *J. Parenteral Enternal Nutr. 18*, 486–90.

Hooge, D. M. (2004), Meta-analysis of broiler chicken pen trials evaluating dietary mannan oligosaccharide, 1993-2003. *Int. J. Poult. Sci. 3*, 163–74.

Horrocks, S. M., Anderson, R. C., Nisbet, D. J. and Ricke, S. C. (2009), Incidence and ecology of *Campylobacter* in animals. *Anaerobe 15*, 18–25.

Howard, Z. R., O'Bryan, C. A., Crandall, P. G. and Ricke, S. C. (2012), *Salmonella* Enteritidis in shell eggs: Current issues and prospects for control. *Food Res. Int. 45*, 755–64.

Hume, M. E. (2011), Historic perspective: Prebiotics, probiotics, and other alternatives to antibiotics. *Poult. Sci. 90*, 2663–9.

Hutkins, R. W., Krumbeck, J. A., Bindels, L. B., Cani, P. D., Fahey, Jr., G. C., Goh, Y. J., Hamaker, B., Martens, E. C., Mills, D. A., Rastal, R. A., Vaughan, E. and Sanders, M. E. (2016), Prebiotics: Why definitions matter. *Curr. Opin. Biotech. 37*, 1–7.

Indikova, I., Humphrey, T. J. and Hilbert, F. (2015), Survival with a helping hand: *Campylobacter* and microbiota. *Front. Microbiol. 6 (Article 1266)*, 1–6. doi: 10.3389/fmicb.2015.01266.

Ishihara, N., Chu, D.-C., Akachi, S. and Juneja, L. R. (2000), Preventive effect of partially hydrolyzed guar gum on infection of *Salmonella enteritidis* in young and laying hens. *Poult. Sci. 79*, 689–97.

Joerger, R. D. (2003), Alternatives to antibiotics: Bacteriocins, antimicrobial peptides and bacteriophages. *Poult. Sci. 82*, 640–7.

Jones, F. T. and Ricke, S. C. (2003), Observations on the history of the development of antimicrobials and their use in poultry feeds. *Poult. Sci. 82*, 613–17.

Jones, G. H. and Ballou, C. E. (1968), Isolation of an α-mannosidase which hydrolyzes yeast mannan – Structure of the backbone of yeast mannan. *J. Biol. Chem. 243*, 2442–6.

Jouany, J. P. (2001), A new look at yeast cultures as probiotics for ruminants. *Feed Mix 9*, 17–19.

Józefiak, D., Rutkowski, A. and Martin, S. A. (2004), Carbohydrate fermentation in the ceca: A review. *Anim. Feed Sci. Technol. 113*, 1–15.

Kaplan, H. and Hutkins, R. W. (2000), Fermentation of fructooligosaccharides by lactic acid bacteria and bifidobacteria. *Appl. Environ. Microbiol. 66*, 2682–4.

Kermanshahi, H. and Rostami, H. (2006), Influence of supplemental dried whey on broiler performance and cecal flora. *Int. J. Poult. Sci. 5*, 538–43.

Klis, F. M., Mol, P., Hellingwerf, K. and Brul, S. (2002), Dynamics of cell wall structure in *Saccharomyces cerevisiae*. *FEMS Microbiol. Revs. 26*, 239–56.

Knudsen, K. E. B. (2014), Fiber and nonstarch polysaccharide content and variation in common crops used in broiler diets. *Poult. Sci. 93*, 2380–93.

Kumar, A., Henderson, A., Forster, G. M., Goodyear, A. W., Weir, T. L., Leach, J. E., Dow, S. W. and Ryan, E. P. (2012), Dietary rice bran promotes resistance to *Salmonella enterica* serovar Typhimurium colonization in mice. *BMC Microbiol. 12*, 71 (http://www.biomedcentral.com/1471–2180/12/71).

Kuriki, T., Yanase, M., Takata, H., Takesada, Y., Imanaka, T. and Okada, S. (1993), A new way of producing isomalto-oligosaccharide syrup by using the transglycosylation reaction of neopullulanase. *Appl. Environ. Microbiol. 59*, 953–9.

Lee, J. T., Bailey, C. A. and Cartwright, A. L. (2003), Guar meal germ and hull fractions differently affect growth performance and intestinal viscosity of broiler chickens. *Poult. Sci. 82*, 1589–95.

Lee, J. T., Connor-Appleton, S., Bailey, C. A. and Cartwright, A. L. (2005), Effects of guar meal by-product with and without β-mannanase hemicell on broiler performance. *Poult. Sci. 84*, 1261–7.

Lee, Y.-C. and Ballou, C. E. (1965), Preparation of mannobiose, mannotriose, and a new mannotetraose from *Saccharomyces cerevisiae* mannan. *Biochemistry 4*, 257–64.

Lianou, A. and Koutsoumanis, K. P. (2013), Strain variability of the behavior of foodborne bacterial pathogens: A review. *Int. J. Food Microbiol. 167*, 310–21.

Lindberg, J. E. (2014), Fiber effects in nutrition and gut health in pigs. *J. Anim. Sci. Biotechnol. 5*, 15 (http://www.jasbsci.com/content/5/1/15).

Luo, P. G., Tzeng, T.-R. J., Qu, L, Lin, Y., Caldwell, E., Latour, R. A., Stutzenberger, F. and Sun, Y. P. (2005), Quantitative analysis of bacterial aggregation mediated by bioactive nanoparticles. *J. Biomed. Nanotech. 2*, 1–5.

Macfarlane, G. T., Steed, H. and Macfarlane, S. (2008), Bacterial metabolism and health-related effects of galacto-oligosaccharides and other prebiotics. *J. Appl. Microbiol. 104*, 305–44.

Mahoney, R. R. (1998), Galactosyl-oligosaccharide formation during lactose hydrolysis: A review. *Food Chem. 63*, 147–54.

Monsan, P. and Paul, F. (1995), Enzymatic synthesis of oligosaccharides. *FEMS Microbiol. Revs 16*, 187–92.

Moran, Jr., E. T. (1985), Digestion and absorption of carbohydrates in fowl and events through perinatal development. *J. Nutr. 115*, 665–74.

Moriceau, S., Besson, C., Levrat, M.-A., Moundras, C., Rémésy, C., Morand, C. and Demigné, C. (2000), Cholesterol-lowering effects of guar gum: Changes in bile acid pools and intestinal reabsorption. *Lipids 35*, 437–44.

Nagpal, M. L. Agrawal, O. P. and Bhatia, I. S. (1971), Chemical and biological examination of guar-meal (*Cyamopsis tetragonoloba* L.). *Ind. J. Anim. Sci. 41*, 283–93.

Naughton, P. J., Mikkelsen, L. L. and Jensen, B. B. (2001), Effects of nondigestible oligosaccharides on *Salmonella enterica* serovar Typhimurium and nonpathogenic *Escherichia coli* in the pig small intestine in vitro. *Appl. Environ. Microbiol. 67*, 3391–5.

Nisbet, D. J. and Martin, S. A. (1991), Effect of a *Saccharomyces cerevisiae* culture on lactate utilization by the ruminal bacterium *Selenomonas ruminantium*. *J. Anim. Sci. 69*, 4628–33.

Nisbet, D. J., Corrier, D. E. and DeLoach, J. R. (1993a), Effect of mixed cecal microflora maintained in continuous culture and of dietary lactose on Salmonella typhimurium colonization in broiler chicks. *Avian Dis. 37*, 528–35.

Nisbet, D. J., Corrier, D. E., Scanlan, C. M., Hollister, A. G., Beier, R. C. and DeLoach, J. R. (1993b), Effect of a defined continuous-flow derived bacterial culture and dietary lactose on Salmonella typhimurium colonization in broiler chickens. *Avian Dis. 37*, 1017–25.

Nisbet, D. J., Ricke, S. C., Scanlan, C. M., Corrier, D. E., Hollister, A. G. and DeLoach, J. R. (1994), Inoculation of broiler chicks with a continuous-flow derived bacterial culture facilitates early cecal bacterial colonization and increases resistance to *Salmonella typhimurium*. *J. Food Prot. 57*, 12–15.

Nisbet, D. J., Corrier, D. E., Ricke, S. C., Hume, M. E., Byrd, J. A. and DeLoach, J. R. (1996a), Maintenance of the biological efficacy in chicks of a cecal competitive-exclusion culture against *Salmonella* by continuous-flow fermentation. *J. Food Prot. 59*, 1279–83.

Nisbet, D. J., Corrier, D. E., Ricke, S. C., Hume, M. E., Byrd, J. A. and DeLoach, J. R. (1996b), Cecal propionic acid as a biological indicator of the early establishment of a microbial ecosystem inhibitory to *Salmonella* in chicks. *Anaerobe, 2*, 345–50.

Nisbet, D. (2002), Defined competitive exclusion cultures in the prevention of enteropathogen colonisation in poultry and swine. *Antonie van Leeuwenhoek. 81*, 481–6.

Nollet, L., Huyghebaert, G. and Spring, P. (2007), Effect of dietary mannan oligosaccharide (Bio-Mos) on live performance of broiler chickens given an anticoccidial vaccine (Paracox) followed by a mild coccidial challenge. *J. Appl. Poult. Res. 16*, 397–403.

O'Bryan, C. A., Pendleton, S. J., Crandall, P. G. and Ricke, S. C. (2015), Potential of plant essential oils and their components in animal agriculture – *in vitro* studies on antibacterial mode of action. *Front. Vet. Sci. 2 (Article 35)*, 1–8. doi: 10.3389/fvets.2015.00035.

Okubo, T., Ishihara, N., Takahashi, H., Fujisawa, T., Kim, M., Yamamoto, T. and Mitsuoka, T. (1994), Effects of partially hydrolyzed guar gum on human intestinal microflora and its metabolism. *Biosci. Biotech. Biochem. 58*, 1364–9.

Ou, S., Kwok, K., Li, Y. and Fu, L. (2001), In vitro study of possible role of dietary fiber in lowering postprandial serum glucose. *J. Agric. Food Chem. 49*, 1026–9.

Oyofo, B. A., DeLoach, J. R., Corrier, D. E., Norman, J. O., Ziprin, R. L. and Mollenhauer, H. H. (1989a), Prevention of *Salmonella typhimurium* colonization of broilers with D-mannose. *Poult. Sci. 68*, 1357–60.

Oyofo, B. A., Droleskey, R. E., Norman, J. O., Mollenhauer, H. H., Ziprin, R. L., Corrier, D. E. and DeLoach, J. R. (1989b), Inhibition by mannose of *in vitro* colonization of chicken small intestine by *Salmonella typhimurium. Poult. Sci. 68*, 1351–6.

Park, S. H., Hanning, I., Perrota, A., Bench, B. J., Alm, E. and Ricke, S. C. (2013), Modifying the gastrointestinal ecology in alternatively raised poultry and the potential for molecular and metabolomic assessment. *Poult. Sci. 92*, 546–61.

Park S. Y., Woodward, C. L., Kubena., L. F., Nisbet., D. J., Birkhold, S. G. and Ricke, S. C. (2008), Environmental dissemination of foodborne *Salmonella* in preharvest poultry production: Reservoirs, critical factors, and research strategies. *Crit. Revs. Environ. Sci. Technol. 38*, 73–111.

Patel, M. B. and McGinnis, J. (1985), The effect of autoclaving and enzyme supplementation of guar meal on the performance of chicks and laying hens. *Poult. Sci. 64*, 1148–56.

Patterson, J. A. and Burkholder, K. (2003), Application of prebiotics and probiotics in poultry production. *Poult. Sci. 82*, 627–31.

Raschke, W. C., Kern, K. A., Antalis, C. and Ballo, C. E. (1973), Genetic control of yeast mannan structure. Isolation and characterization of mannan mutants. *J. Biol. Chem. 248*, 4660–6.

Rastall, R. (2004), Bacteria in the gut: Friends and foes and how to alter the balance. *J. Nutr. 134*, 2022S–6S.

Rastall, R. A. and Maitin, V. (2002), Prebiotics and synbiotics: Towards the next generation. *Curr. Opin. Biotechnol. 13*, 490–6.

Rastall, R. A., Gibson, G. R., Gill, H. S., Guarner, F., Klaenhammer, T. R., Pot, B., Reid, G., Rowland, I. R. and Sanders, M. E. (2005), Modulation of the microbial ecology of the human colon by probiotics and synbiotics to enhance human health: An overview of enabling science and potential applications. *FEMS Microbiol. Ecology 52*, 145–52.

Revolledo, L., Ferreira, A. J. P. and Mead, G. C. (2006), Prospects in *Salmonella* control: Competitive exclusion, probiotics, and enhancement of avian intestinal immunity. *J. Appl. Poult. Res. 15*, 341–51.

Ricke, S. C., Martin, S. A. and Nisbet, D. J. (1996), Ecology, metabolism, and genetics of ruminal selenomonads. *Crit. Revs. Microbiol. 22*, 27–65.

Ricke, S. C. and Pillai, S. D. (1999), Conventional and molecular methods for understanding probiotic bacteria functionality in gastrointestinal tracts. *Crit. Revs. Microbiol. 25*, 19–38.

Ricke, S. C. (2003a), Perspectives on the use of organic acids and short chain fatty acids as antimicrobials. *Poult. Sci. 82*, 632–9.

Ricke, S. C. (2003b), The gastrointestinal tract ecology of *Salmonella* Enteritidis colonization in molting hens. *Poult. Sci. 82*, 1003–7.

Ricke, S. C., Kundinger, M. M., Miller, D. R. and Keeton, J. T. (2005), Alternatives to antibiotics: Chemical and physical antimicrobial interventions and foodborne pathogen response. *Poult. Sci. 84*, 667–75.

Ricke, S. C., Hererra, P. and Biswas, D. (2012a), Chapter 23. *Bacteriophages for potential food safety applications in organic meat production*. In S. C. Ricke, E. J. Van Loo, M. G. Johnson and C. A. O' Bryan (eds), *Organic Meat Production and Processing*. Wiley Scientific/IFT, New York, NY, pp. 407–24.

Ricke, S. C., Van Loo, E. J., Johnson M. G. and O'Bryan, C. A. (eds) (2012b), *Organic Meat Production and Processing*. Wiley Scientific/IFT, New York, NY.

Ricke, S. C., Dunkley, C. S. and Durant, J. A. (2013), A review on development of novel strategies for controlling *Salmonella* Enteritidis colonization in laying hens: Fiber – based molt diets. *Poult. Sci. 92*, 502–25.

Ricke, S. C. and Hanning, I. B. (2013), Chapter 9. *Food safety applications of nanoparticles*. In R. Asmatulu (ed.), *Nanotechnology Safety*. Elsevier, Amsterdam, The Netherlands, pp. 115–25.

Ricke, S. C. (2014), Application of molecular approaches for understanding foodborne *Salmonella* establishment in poultry production. *Adv. Biol. Vol.* Article ID 813275, p. 25. doi:10.1155/2014/813275.

Ricke, S. C. (2015), Potential of fructooligosaccharide prebiotics in alternative and nonconventional poultry production systems. *Poult. Sci. 94*, 1411–18.

Ricke, S. C. and Saengkerdsub, S. (2015), Chapter 19. *Bacillus probiotics and biologicals for improving animal and human health: Current applications and future prospects*. In V. Ravishankar Rai and A. Jamuna Bai (eds), *Beneficial Microbes in Fermented and Functional Foods*. CRC Press/Taylor & Francis Group, Boca Raton, FL, pp. 341–60.

Rivera Calo, J., Crandall, P. G., O'Bryan, C. A. and Ricke, S. C. (2015), Essential oils as antimicrobials in food systems – A review. *Food Control. 54*, 111–19.

Roberfroid, M. B. (2007), Prebiotics: The concept revisited. *J. Nutr. 137*, 830S–7S.

Roberfroid, M. B. (2000), Prebiotics and probiotics: Are they functional foods? *Am J Clin Nutr. 71*, 1682–7.

Rossi, M., Corradini, C., Amaretti, A., Nicolini, M., Pompei, A., Azanoni, S. and Matteuzzi, D. (2005), Fermentation of fructooligosaccharides and inulin by bifidobacteria: A comparative study of pure and fecal cultures. *Appl. Environ. Microbiol. 71*, 6150–8.

Roto, S. M., Rubinelli, P. M. and Ricke, S. C. (2015), An introduction to the avian gut microbiota and the effects of yeast-based prebiotic compounds as potential feed additives. *Front. Vet. Sci. 2 (Article 28)*, 1–18. doi: 10.3389/fvets.2015.00028.

Saminathan, M., Sieo, C. C., Kalavathy, R., Abdullah, N. and Ho, Y. W. (2011), Effect of prebiotic oligosaccharides on growth of *Lactobacillus* strains used as a probiotic for chickens. *African J. Microbiol. Res. 5(1)*, 57–64.

Schell, M. A., Karmirantzou, M., Snel, B., Vilanova, D., Berger, B., Pessi, G., Zwahlen, M.-C., Desiere, F., Bork, P., Delley, M., Pridmore, R. D. and Arigoni, F. (2002), The genome sequence of *Bifidobacterium longum* reflects its adaptation to the human gastrointestinal tract. *PNAS 99*, 14422–7.

Schrezenmeir, J. and de Vrese, M. (2001), Probiotics, prebiotics, and synbiotics – approaching a definition. *Am. J. Clin. Nutr. 73*, 361S–4S.

Sheng, K.-C., Pouniotis, D. S., Wright, M. D., Tang, C. K., Lazoura, E., Pietersz, G. A. and Apostolopoulos, V. (2006), Mannan derivatives induce phenotypic and functional maturation of mouse dendritic cells. *Immunology 118*, 372–83.

Shoaf, K., Mulvey, G. L., Armstrong, G. D. and Hutkins, R. W. (2006), Prebiotic galactooligosaccharides reduce adherence of enteropathogenic *Escherichia coli* to tissue culture cells. *Infect. Immun. 74*, 6920–8.

Siragusa, G. R. and Ricke, S. C. (2012), Chapter 20. *Probiotics as pathogen control agents for organic meat production*. In S. C. Ricke, E. J. Van Loo, M. G. Johnson and C. A. O' Bryan (eds), *Organic Meat Production and Processing*. Wiley Scientific/IFT, New York, NY, pp. 331–49.

Slavin, J. L. and Greenberg, N. A. (2003), Partially hydrolyzed guar gum: Clinical nutrition uses. *Nutrition 19*, 549–52.

Sofos, J. N., Flick, G., Nychas, G.-J., O'Bryan, C. A., Ricke, S. C. and Crandall, P. G. (2013), Chapter 6. *Meat, Poultry, and Seafood.* In M. P. Doyle and R. L. Buchanan (eds), *Food Microbiology-Fundamentals and Frontiers, 4th Edition.* American Society for Microbiology, Washington, D.C., pp. 111–67.

Spring, P., Wenk, C., Dawson, K. A. and Newman, K. E. (2000), The effects of dietary mannanoligosaccharides on cecal parameters and the concentrations of enteric bacteria in the ceca of *Salmonella*-challenged broiler chicks. *Poult. Sci. 79*, 205–11.

Stavric S. (1992), Defined cultures and prospects. *Int. J. Food Microbiol. 15*, 245–63.

Steer, T., Carpenter, H., Tuohy, K. and Gibson, G. R. (2000), Perspectives on the role of the human gut microbiota and its modulation by pro and prebiotics. *Nutr. Res. Rev. 13*, 229–54.

Stewart, T. S., Mendershausen, P. B. and Ballou, C. E. (1968), Preparation of a mannopentaose, mannohexaose, and mannoheptaose from *Saccharomyces cerevisiae* mannan. *Biochemistry 7*, 1843–54.

Taylor, S., Qu, L., Kitaygorodskiy, A., Teske, J., Latour, R. A. and Sun, Y. P. (2004), Synthesis and characterization of peptide-functionalized polymeric nanoparticles. *Biomacromolecules 5*, 245–8.

Ten Bruggencate, S. J. M., Bovee-Oudenhoven, I. M. J., Lettink-Wissink, M. L. G. and Van der Meer, R. (2003), Dietary fructo-oligosaccharides dose-dependently increase translocation of *Salmonella* in rats. *J. Nutr. 133*, 2313–18.

Thanissery, R., McReynolds, J. L., Conner, D. E., Macklin, K. S., Curtis, P. A. and Fasina, Y. O. (2010), Evaluation of the efficacy of NuPro yeast extract in reducing intestinal *Clostridium perfringens* levels in broiler chickens. *Poult. Sci., 89*, 2380–8.

Thitaram, S. N., Chung, C.-H., Day, D. F., Hinton Jr., A., Bailey, J. S. and Siragusa, G. R., (2005), Isomaltooligosaccharide increases cecal bifidobacteria population in young broiler chickens. *Poult. Sci. 84*, 998–1003.

Tizard, I. R., Carpenter, R. H., McAnalley, B. H. and Kemp, M. C. (1989), The biological activities of mannans and related complex carbohydrates. *Mol. Biother. 1*, 290–6.

Tuohy, K. M., Kolida, S., Lustenberger, A. M. and Gibson, G. R. (2001), The prebiotic effects of biscuits containing partially hydrolysed guar gum and fructo-oligosaccharides – A human volunteer study. *Br. J. Nutr. 86*, 341–8.

Tzortzis, G., Goulas, A. K. and Gibson, G. R. (2005), Synthesis of prebiotic galactooligosaccharides using whole cells of a novel strain, *Bifidobacterium bifidum* NCIMB 41171. *Appl. Micobiol. Biotech. 68*, 412–16.

Vandeplas, S., Dauphin, R. D., Beckers, Y., Thonart, P. and Thévis, A. (2010), *Salmonella* in chicken: Current and developing strategies to reduce contamination at farm level. *J. Food Prot. 73*, 774–85.

Van Etten, C. H., Miller, R. W., Wolff, I. A. and Jones, Q. (1961), Amino acid composition of twenty-seven selected seed meals. *Agric. Food Chem. 9*, 79–82.

Verdonk, J. M. A. J., Shim, S. B., Van Leeuwen, P. and Verstegen, M. W. A. (2005), Application of inulin-type fructans in animal feed and pet food. *Br. J. Nutr. 93*, S125–38.

Verma, S. V. S. and MacNab, J. M. (1982), Guar meal in diets for broiler chickens. *Br. Poult. Sci. 23*, 95–105.

Verma, S. V. S. and MacNab, J. M. (1984), Chemical, biochemical and microbiological examination of guar meal. *Ind. J. Poult. Sci. 19*, 165–70.

Vohra, P. and Kratzer, F. H. (1964a), The use of guar meal in chicken rations. *Poult. Sci. 43*, 502–3.

Vohra, P. and Kratzer, F. H. (1964b), Growth inbitory effect of certain polysaccharides for chickens. *Poult. Sci. 43*, 1164–70.

Waldroup, A. L., Yamaguchi, W., Skinner, J. T. and Waldroup, P. W. (1992), Effects of dietary lactose on incidence and levels of salomonellae on carcasses of broiler chickens grown to market age. *Poult. Sci. 71*, 288–95.

Wang, Y. (2009), Prebiotics: Present and future science and technology. *Food Res. Int. 42*, 8–12.

Xu, Z. R., Hu, C. H., Xia, M. S., Zhan, X. A. and Wang, M. Q. (2003), Effects of dietary fructooligosaccharide on digestive enzyme activities, intestinal microflora and morphology of male broilers. *Poult. Sci. 82*, 1030–6.

Yamada, K., Tokunaga, Y., Ikeda, A., Ohkura, K., Mamiya, S., Kaku, S., Sugano, M. and Tachibana, H. (1999), Dietary effect of guar gum and its partially hydrolyzed product on the lipid metabolism and immune function of Sprague-Dawley rats. *Biosci. Biotech. Biochem. 63*, 2163–7.

Yamamoto, Y., Sogawa, I., Nishina, A., Saeki, S., Ichikawa, N. and Iibata, S. (2000), Improved hypolipidemic effects of xanthan gum-galactomannan. *Biosci. Biotech. Biochem. 64*, 2165–71.

Yang, Y., Iji, P. A., Kocher, A., Mikkelsen, L. L. and Choct, M. (2007), Effects of mannanoligosaccharide on growth performance, the development of gut microflora, and gut function of broiler chickens raised on new litter. *J. Appl. Poult. Res. 16*, 280–8.

Yang, Y., Iji, P. A., Kocher, A., Thomsen, E., Mikkelsen, L. L. and Choct, M. (2008a), Effects of mannanoligosaccharide in broiler chicken diets on growth performance, energy utilisation, nutrient digestibility and intestinal microflora. *Br. Poult. Sci. 49*, 186–94.

Yang, Y., Iji, P. A., Kocher, A., Mikkelsen, L. L. and Choct, M. (2008b), Effects of mannanoligosaccharide and fructooligosaccharide on the response of broilers to pathogenic *Escherichia coli* challenge. *Br. Poult. Sci. 49*, 550–9.

Yang, Y., Iji, P. A. and Choct, M. (2009), Dietary modulation of gut microflora in broiler chickens: A review of the role of six kinds of alternatives to in-feed antibiotics. *World's Poult. Sci. J. 65*, 97–114.

Yoon, S.-J., Chu, D.-C. and Juneja, L. R. (2008), Chemical and physical properties, safety and application of partially hydrolized guar gum as dietary fiber. *J. Clin. Biochem. Nutr. 42*, 1–7.

Yusrizal, X. N. and Chen, T. C. (2003), Effect of adding chicory fructans in feed on fecal and intestinal microflora and excreta volatile ammonia. *Int. J. Poult. Sci. 2*, 188–94.

Zhang, W. F., Li, D. F., Lu, W Q. and Yi, G. F. (2003), Effects of isomalto-oligosaccharides on broiler performance and intestinal microflora. *Poult. Sci. 82*, 657–63.

Zimmermann, N. G., Andrews, D. K. and McGinnis, J. (1987), Comparison of several induced molting methods on subsequent performance of single comb white leghorn hens. *Poult. Sci. 66*, 408–17.

Ziprin, R. L., Corrier, D. E., Hinton, Jr., A., Beier, R. C., Spates, G. E. and DeLoach, J. R. (1990), Intracloacal Salmonella typhimurium infection of broiler chickens: Reduction of colonization with anaerobic organisms and dietary lactose. *Avian Dis. 34*, 749–53.

9 781801 464208